ECONOMIC THICKNESS FOR INDUSTRIAL INSULATION

ECONOMIC THICKNESS FOR INDUSTRIAL INSULATION

Prepared by

THE U.S. DEPARTMENT OF ENERGY
Office of Industrial Programs

Performed under Contract No. CO-04-50169

ECONOMIC THICKNESS FOR INDUSTRIAL INSULATION

This report was prepared as an account of work sponsored by an agency of the United States Government. Neither the United States Government nor any agency thereof, nor any of their employees, contractors, subcontractors, or their employees, makes any warranty, express or implied, nor assumes any legal liability or responsibility for any third party's use or the results of such use of any information, apparatus, product or process disclosed in this report, nor represents that its use by such third party would not infringe privately owned rights.

Any views or recommendations contained in this report do not necessarily reflect the views of the U.S. Department of Energy, nor does mention of trade names or commercial products represent endorsement or recommendation for use by D.O.E.

Library of Congress Catalog Card No. 83-080102
ISBN: 0-915586-72-X

Published and reprinted in 1983 by:
The Fairmont Press Inc.
P.O. Box 14227
Atlanta, GA 30324

Distributed in Europe, Japan, India, Middle East, Southeast Asia, Africa and the West Indies by E. & F.N. Spon, II New Fetter Lane, London, EC4P4EE

ACKNOWLEDGEMENTS

Appreciation is expressed to the following persons and organizations for the information and assistance they provided in developing this manual.

- J. Barnhart, Thermal Insulation Manufacturers Association
- G. Barnett, Stone & Webster, Inc.
- R. Gussman, Energy Research and Development Association
- J. Johnson, Shell Oil Corp.
- V. Lockary and R. Davis, Bechtel Corp.
- J. McAllister, American Electric Power
- H. Moak, E. I. Dupont de Nemours and Co.
- R. Nurf, Exxon Chemical Co.
- E. Powell, Union Carbide Corp.
- M. Reece, Fluor Engineers and Constructors, Inc.
- D. Roem, M. W. Kellogg Co.
- C. Smolenski, Pittsburgh Corning Corp.
- V. Toepp, Texaco, Inc.
- W. Wartan, Sargent and Lundy Engineers, Inc.
- P. Weaver, Dow Chemical Corp.

A special note of thanks is extended to the Energy Task Force Committee of the National Insulation Contractors Association for their helpful and time consuming efforts in the insulation cost portion of this manual.

- W. R. Murfin, B & B Insulation, Chairman
- R. Anderson, Anco Insulations, Inc.
- T. Brodie, New England Insulation Co.
- B. Crockett, Willard A. Selle, Inc.
- J. Crumit, Campbell McCormick Co.
- I. Dewar, Dewar Insulations, Inc.
- C. Fowler, ACANDS
- D. Healey, Owens-Corning Fiberglass Corp.
- A. Keasbey, Robert A. Keasbey Co.
- W. Killion, Shook & Fletcher Insulation Co.
- J. Koach, N.I.C.A.
- D. Luse, Luse Stevenson Co.
- W. Morrow, All-Fields Insulation, Inc.
- H. Mullenix, North Brothers Co.
- J. Royer, Young Sales Corp.
- L. Saxby, Owens-Corning Fiberglas Corp.
- F. Vite, Johns-Manville Sales Corp.
- R. Stapleton, Johns-Manville Sales Corp.
- L. Sweetser, Metalclad Insulation Corp.
- C. Wheatley, Pacor, Inc.
- R. Wopperer, Frontier Insulation Contractors

CONTENTS

CHAPTER 1:	INTRODUCTION	1
CHAPTER 2:	THE ECONOMICS OF INSULATION	3
	Introduction	3
	Economic Analysis	3
	Cost of Lost Energy	4
	Cost of Insulation	5
	Theory	5
	Life Cycle Costing	8
	Present Value Analysis	9
CHAPTER 3:	COST OF ENERGY	10
	Contents	10
	Glossary	10
	Choice of Insulation and Facility Life Terms, n_1, i_2, and i_3	12
	Choice of Cost of Money Terms i_1, i_2, and i_3	13
	Cost of Heat Worksheet	14
	Cost of Refrigeration Worksheet	16
CHAPTER 4:	THE COST OF INSULATION	34
	Introduction	34
	Glossary	35
	Determining Incremental Insulation Costs	35
	Insulation Cost Worksheet	37
	Estimating Price per Unit of Insulation	39
	Material Prices	39
	Labor for Installation	40
	Installed Cost Estimating Procedure	41
	Insulation Cost Estimator Worksheet	43
CHAPTER 5:	ECONOMIC THICKNESS DETERMINATION	50
	Introduction	50
	Contents	50
	Glossary	50
	Economic Thickness Determination Worksheet	52
CHAPTER 6:	CONDENSATION CONTROL	81
	Introduction	81
	The Condensation Problem	81
	Determining Minimum Insulation Thickness	82
	Methodology	83
	Condensation Control Worksheet	85

CHAPTER 7:	RETROFITTING INSULATION	96
	Introduction	96
	Procedure	96
	Glossary	96
	Retrofit Insulation Worksheet	98
CHAPTER 8:	SAMPLE PROBLEMS	100
	No. 1 — High Temperature Steam Pipe	100
	No. 2 — Low Temperature Steam Pipe and Chilled Water Line	121
	No. 3 — Retrofit Condensate Line/Payback Analysis	131
BIBLIOGRAPHY		136
APPENDIX A:	DERIVATION OF EQUATIONS	139
For Pipe		139
Flat Plate		145
Parameters		147
APPENDIX B:	SENSITIVITY ANALYSIS	156
APPENDIX C:	COST OF INSTALLED INSULATION	171
APPENDIX D:	THICKNESS TO PREVENT CONDENSATION	176
APPENDIX E:	RETROFIT — MULTILAYER INSULATION	182
APPENDIX F:	PIPING COMPLEXITY FACTORS	188

TABLES

3-1:	S Values for Refrigeration Systems	18
3-2:	Typical Heating Values of Various Coals, Btu/lb x 10^3	19
3-3:	Saturated Steam Enthalpy Values	20
4-1:	Piping Complexity Factors, PC	38
4-2:	Unit Selling Price Ratios	44
4-3:	Unit Material Selling Price Ratios — Performed Mineral Wool	45
4-4:	Unit Material Selling Price Ratios — Performed Fiber Glass	46
4-5:	Unit Material Selling Price Ratios — Cellular Glass	47
4-6:	Material Type Correction Factors, F_t	48
4-7:	Regional Labor Productivity Factors, F_r	48
4-8:	Base Worker Productivity, WP	49
6-1:	Design Summer Conditions for the United States	86-88
6-2:	Table of Relative Humidities	89
6-3:	Emittance Ranges for Metal Jackets, Mastics, and Various Surface Finishes	90
F-1:	Average Number of Fittings per 100 Linear Feet Pipe For Welded Piping	189
F-2:	Equivalent Length Cost Factor for Pipe Fittings	190
F-3:	Approximate Outside Surface Areas of Fitting Insulation (1½-inch thickness) Expressed in Equivalent feet of Pipe Insulation	191

ILLUSTRATIONS

2-1:	Insulation Thickness vs. Insulation Cost	6
3-1:	Multiplier to Apply to Present Costs for Determining the Average Annual Costs when Uniform Cost Increases Occur in Future Years	21
3-2a:	The Cost of Heat, C_h, for Fuel Oil as the Heat Source	22
3-2b:	The Cost of Heat, C_h, for Gas as the Heat Source	23
3-2c:	The Cost of Heat, C_h, for Coal as the Heat Source	24
3-2d:	The Cost of Heat, C_h, for Electricity as the Heat Source	25
3-3:	The Average Annual Value of Heat Cost Including Operation and Maintenance Costs at the Heat Producing Facility	26
3-4:	Compound Interest Factor for Use with Figure 3-5	27
3-5:	The Amortization Multiplier, B	28
3-6:	The Distribution of the Heat Production Capital Costs over the Energy Output on an Annual Basis	29
3-7:	Coefficient of Performance of Refrigeration Systems	30
3-8:	The Operating Cost of Refrigeration Systems	31
3-9:	Average Make-up Water Cost of Water Cooled Refrigeration System	32
3-10:	The Annual Average Cost of Capital Investment in a Refrigeration System	33
5-1:	High Temperature Insulation Materials, k Factors	34
5-2:	Low Temperature Insulation Materials, k Factors	54
5-3:	Annual Cost of Heat Lost or Gained — D_p, D_{pr}, D_s, and D_{sr} Values	55
5-4:	Compound Interest Factor for Use with Figure 5-5	56
5-5:	The Amortizing Multiplier, B	57
5-6:	Z_s, Factor for Flat Surfaces	58
5-7:	Z_p, Factor for Round (Pipe) Surface	59
5-8a:	Economic Thickness, Flat Surface	60
5-8b:	Economic Thickness, ½-inch Pipe	61
5-8c:	Economic Thickness, ¾-inch Pipe	62
5-8d:	Economic Thickness, 1-inch Pipe	63
5-8e:	Economic Thickness, 1½-inch Pipe	64
5-8f:	Economic Thickness, 2-inch Pipe	65
5-8g:	Economic Thickness, 2½-inch Pipe	66
5-8h:	Economic Thickness, 3-inch Pipe	67
5-8i:	Economic Thickness, 4-inch Pipe	68
5-8j:	Economic Thickness, 5-inch Pipe	69
5-8k:	Economic Thickness, 6-inch Pipe	70
5-8l:	Economic Thickness, 8-inch Pipe	71
5-8m:	Economic Thickness, 10-inch Pipe	72

5-8n:	Economic Thickness, 12-inch Pipe	73
5-8o:	Economic Thickness, 14-inch Pipe	74
5-8p:	Economic Thickness, 16-inch Pipe	75
5-8q:	Economic Thickness, 18-inch Pipe	76
5-8r:	Economic Thickness, 20-inch Pipe	77
5-8s:	Economic Thickness, 24-inch Pipe	78
5-8t:	Economic Thickness, 30-inch Pipe	79
5-8u:	Economic Thickness, 36-inch Pipe	80
6-1:	Dewpoint from Relative Humidity and Dry Bulb Temperature	91
6-2:	Radiation Heat Flow from Surface of Insulation	92
6-3:	Convection Heat Flow from Surface of Insulation	93
6-4:	Equivalent Insulation Thickness for Condensation Prevention	94
6-5:	Actual Insulation Thickness Required to Prevent Condensation	95
A-1:	Basic Economic Thickness Curves	143
A-2:	Heat Transmission vs. Surface Resistance, Flat and Cylindrical Surface	148
B-1:	Computer Readout for Sensitivity Analysis of Various Parameters	157
B-2:	Sensitivity Analysis of Parameter k in Equation B-1	158
B-3:	Sensitivity Analysis of Parameters Y and T in Equation B-1	159
B-4:	Sensitivity Analysis of Parameter R_s in Equation B-1	160
B-5:	Sensitivity Analysis of Parameter m_c in Equation B-1	161
B-6:	Sensitivity Analysis of Parameter i_3 in Equation B-1	162
B-7:	Sensitivity Analysis of Parameter i_1 in Equation B-1	163
B-8:	Sensitivity Analysis of Parameter n_1 in Equation B-1	164
B-9:	Sensitivity Analysis of Parameter C in Equation B-1	165
B-10:	Sensitivity Analysis of Parmeters E and H in Equation B-1	166
B-11:	Sensitivity Anlaysis of Parameter P in Equation B-1	167
B-12:	Sensitivity Anlaysis of Parameter Q in Equation B-1	168
B-13:	Sensitivity Analysis of Parameter i_2 in Equation B-1	169
B-14:	Sensitivity Analysis of Parameter n_2 in Equation B-1	170
C-1:	Price of Installed Insulation vs. Insulation Thickness, Plot of Actual Data with Various Means of Fitting Data Points	174
C-2:	Installed Cost of Insulation vs. Insulation Thickness, Data Requirements for m_c Determination	175
D-1:	Parameter Sensitivity for L, Insulation Thickness To Prevent Condensation	180
D-2:	Sample Computer Run of Condensation Control Sensitivity Analysis	181
E-1:	Pipe with Two Insulation Layers	182
E-2:	Means of Finding R_3	184
E-3:	Flat Surface with Two Insulation Layers	185
E-4:	Solution for Least Cost New Insulation Layer Over Existing Layer	186

CHAPTER 1
INTRODUCTION

Industrial plants and utilities account for about half of the total United States energy use. Historically, this sector has utilized thermal insulation to protect personnel, maintain process temperatures and conserve energy. In this period of rising fuel costs, thermal insulation is probably the best proven and universally applicable solution available to industry for the conservation of this expensive commodity.

In past years, when first-cost analysis was the primary criterion for capital expenditures, using more than minimal amounts of insulation was not justified. This rationale is no longer acceptable in the present era of expensive energy. Consequently life cycle costing, especially in areas of energy conservation, is now becoming the predominant industrial design criterion. Insulation specified around today's economics saves on the average 30 to 40 percent more energy than is being conserved with the outdated insulation designs of the past. This is not only a significant factor in new plant construction, but also provides the financial justification for retrofitting insulation in existing facilities.

The conservation of energy through the use of optimal economic insulation thickness has obvious benefits for industry, and equally impressive potential benefits for the United States. This study estimates that over 1400 trillion Btu's, the equivalent of over 122 million barrels of oil could have been saved in 1974, had industry alone installed economic insulation thicknesses. This is in addition to the energy presently being saved with existing insulation. The potential additional energy conservation available to industry through the use of economic insulation through 1990 is estimated to be the equivalent of 3.5 billion barrels of oil or 250 million barrels per year. An extra benefit to society is a reduction in air pollution, which necessarily follows decreased fuel use.

As with any capital expenditure, the dollars spent for insulation are expected to provide the plant owner with a certain return on his investment. (The return can be in the form of personnel and equipment protection as well as in energy saving.) This manual allows the user a large degree of flexibility in designing an insulation system around the particular economic and financial criteria for the plant in question. By estimates made of money costs and insulation life terms, the least-cost insulation thickness can be specified based on minimum life-cycle cost, required percentage return on investment using discounted cash flow or discounted payback period.

All significant technical, as well as economic, inputs affecting the thickness are variable, lending the manual to application for the large majority of pipe and tank insulation design situations. By using nomographs as the means of calculation, the user is spared tedious mathematical computations, although the equations solved with the nomographs are presented for the user's information. A complex appendix in which all equations and procedures are derived has been included.

This manual provides the solution for economic thickness on both hot and cold systems. Since the insulation required to prevent condensate formation on cold systems may exceed the economic specification, a method for calculating the thickness required to prevent condensation has also been included.

A procedure for calculating the economic thickness of insulation retrofitted in existing facilities is presented, as well as example problems on the ecnomic thickness determination.

Manual Use

This manual is presented in 8 chapters plus appendices. Chapters 1 and 2 are included as background information.

The economic insulation thickness is found using the following chapters:

- Chapter 3 Cost of Energy (Heat and Refrigeration)
- Chapter 4 Cost of Insulation
- Chapter 5 Economic Thickness Determination

Each of these chapters contains a worksheet, which provides a step-by-step procedure for solving the problem. Nomographs are included in each section for simplifying the required calcuations. It is suggested that the user make copies of the worksheets, enabling each problem to be recorded and saved for future reference.

For subambient temperature systems, the economic insulation thickness calculated with Chapters 3-5 should be compared to the thickness required to prevent condensation (sweating), as calculated with Chapter 6.

Chapter 7 presents the procedure for calculating economic insulation for retrofit situations. Retrofitting encompasses both the adding of insulation over existing insulation, or insulating bare equipment in an existing facility.

Sample problems are included in Chapter 8, and it is suggested that the user follow the procedures used to solve these sample problems to familiarize himself with the manual method.

Finally, the appendices are provided to allow the user to perform the procedures and equations used in the manual.

CHAPTER 2
THE ECONOMICS OF INSULATION

Introduction

In determining the most economic design for an insulation system, two or more insulating materials may be evaluated for least cost for a given thermal performance; or, optimum insulation thickness may be selected for a specific insulation type. In either case, the decision should be based on which design will save the greatest number of dollars over a specified period, in both initial and continuing costs.

The basic method used to determine economic insulation thickness for a one-material system is presented in this section. This method is applied in the subsequent sections for the solution of economic thickness insulation.

Economic Analysis

The primary function of insulation is to reduce the loss of energy from a surface operating at a temperature other than ambient. The economic use of insulation

1. reduces plant operating expenditures for fuel, power, etc.;
2. improves process efficiency;
3. increases system output capacity or may reduce the required capital cost.

There are two costs associated with the insulation type chosen. For any given thickness, there is

1. a cost for the insulation itself;
2. a cost for the energy lost through this thickness.

The total cost for a given period is the sum of both costs.

The optimum economic thickness is that which provides the most cost effective solution for insulating and is determined when total costs are a minimum. Since the solution calls for the sum of the lost energy and insulation investment costs, both costs must be compared in similar terms. Either the cost of insulation must be estimated for each year and compared to the average annual cost of lost energy over the expected life of the insulation, or the cost of the expected energy loss each year must be expressed in present dollars and compared with the total cost of the insulation investment. The former method, making an annual estimate of the insulation cost and comparing it to the average expected annual cost of lost energy is the method used in this analysis.

Cost of Lost Energy

The rate of energy transfer through the insulation, the cost or value affixed to that energy, and the operational hours per year determine the cost of lost energy per annum. The rate of the energy transfer is a function of the following:

1. the temperature difference across the insulation,
2. the thermal conductivity,
3. the thickness,
4. the thermal resistance of the external surface of the insulation.

These elements are treated in Chapter 5.

The value of energy in a system is directly related to the end use of that energy. For example, high-energy steam capable of driving a turbo-generator has a greater economic value per Btu to a utility than low-energy steam, which cannot be utilized to produce electricity. However, since useful energy has so many end uses, it is not possible for this manual to offer a practical method for calculating energy value based on end use. Another method for determining the value of energy is to add the costs of producing the energy, which is the method used in this manual.

The cost of purchasing or producing energy includes

1. fuel or power costs,
2. capital equipment costs,
3. operating and maintenance expenditures.

Fuel or power costs have the greatest effect upon the dollar value of energy. The efficiency with which these fuels are converted into process heat affects the value assigned to energy in a process. The energy conversion efficiency or coefficient of performance (refrigeration process) of the equipment must be considered. Since fuel and power costs will most likely change with time, the average cost of each over the life of the insulation project is used rather than today's costs.

The energy being conserved because of insulation requires capital equipment for its production. The capital cost of the energy plant must, therefore, be assigned to the dollar cost of energy. This cost is estimated per annum considering the plant depreciation period, the average annual energy production and the cost of money.* Annual maintenance and operating expenses also contribute to the cost of energy, and must be included.

*The cost of money, or discount rate, is the cost of financing the investment.

Cost of Insulation

The cost of insulation is the sum of the annual insulation investment cost and the yearly insulation maintenance expense. The initial cost of the insulation system is prorated over the chosen project life using the appropriate cost of money or required rate of return on the last increment of insulation applied.

The period over which the insulation investment cost is considered is a factor in selecting the economic thickness. If the chosen period is short, the annual insulation cost will be high, the economic thickness will be small, and the insulation system will not provide the lowest total annual cost over the service life of the insulation. It is therefore recommended that the insulation service life be used for the project period if user project payout periods do not otherwise dictate.

Theory

Optimal economic insulation thickness may be arrived at by two methods:

1. The minimum total cost method;
2. The incremental (or marginal) cost method.

For a given situation both methods will yield the same thickness solution.

The minimum total cost method involves the actual calculation of lost energy and insulation costs for each insulation thickness. The thickness producing the lowest total cost is the optimal economic solution. However, the numerous calculations involved in this method require a computer for derivation of an answer; this precludes the use of the minimum total cost method in this manual.

The incremental or marginal cost method provides a simplified and direct solution for the least cost thickness without having to calculate total annual costs. With this method, the optimum thickness is determined to be the point where the last dollar invested in insulation results in exactly $1 in energy cost savings, on a discounted cash flow basis.

The "incremental cost" is a term applied to the change in installed cost between two successive insulation thicknesses. At thickness L, the cost for adding the insulation thickness "ΔL", is given as

$$m_c = \frac{\Delta C}{\Delta L} \quad \text{(See Figure 2-1)}$$

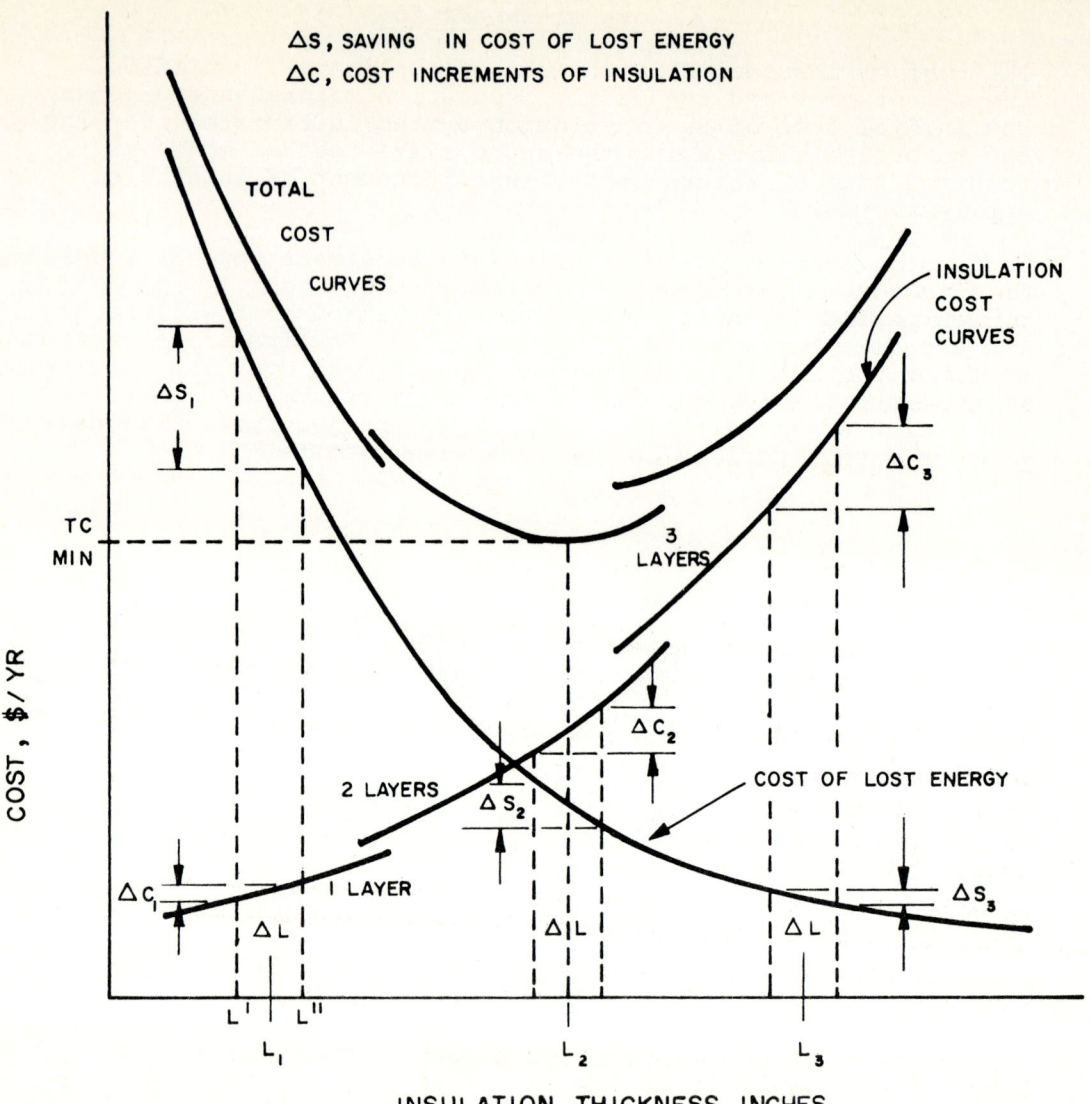

Figure 2-1.   Insulation thickness vs. insulation cost

where

m_c = incremental insulation cost ($/inch)

ΔC = difference in insulation installed cost for thickness L' and L" ($)

ΔL = additional insulation thickness, L" - L' (inches).

The dollar investment required to increase the insulation thickness from L' to L" is therefore determined as m_c.

At thickness L, the reduction in the cost of lost energy obtained by adding the insulation thickness ΔL, is given as

$$m_s = \frac{\Delta S}{\Delta L}$$

where

m_s = incremental savings in the cost of lost energy ($/inch)

ΔS = difference in the cost of lost energy between thicknesses L' and L" ($).

The dollar savings in energy obtained by increasing the insulation thickness from L' to L" is therefore determined as m_S, which always has a negative value because adding insulation reduces lost energy cost.

The total annual cost is equal to the sum of the cost of lost energy and the cost of insulation. The change in total costs when additional insulation is added is equal to the sum of m_S and m_C.

When total cost is minimum, the change in the total cost is equal to zero and m_S equals m_C. When this condition is met, adding more insulation is not profitable because the incremental cost for additional insulation, m_C, becomes greater than the dollar value of the energy saved, m_S, by that additional insulation.

Figure 2-1 depicts both the minimum total cost and incremental methods and reflects the requirement for multiple layers needed in higher insulation thicknesses and/or needed to alleviate expansion and contraction forces in cyclical temperature systems.

Figure 2-1 also depicts typical insulation cost curves for single, double, and triple layers of thickness. As shown, increments of insulation thickness cost more as total thickness increases either within a single layer or as a result of multiple layering. The average slope of the triple layer cost curve is greater than the average slope of double layer cost curve. The reason is that both material and labor costs increase with thickness and with each additional layer. The step between the respective insulation cost curves is primarily the labor cost required to add the additional layer.

The total cost curves are for single, double, and triple layers, and are generated by adding the insulation and lost energy cost curves. For the example shown in Figure 2-1, the optimal economic thickness falls in the double layer range, as total costs are a minimum at point L_2. The insulation thickness, L_2, is the economic solution using the minimum total cost method.

The incremental method is demonstrated in Figure 2-1 by comparing the change in insulation cost (ΔC) with the savings in cost of lost energy (ΔS) resulting from the addition of an insulation thickness (ΔL). At thickness L_1, the addition of ΔL causes a large saving in lost energy (ΔS_1) at a small increase in insulation cost (ΔC_1). At thickness L_3, the addition of ΔL is rather expensive (ΔC_3) and saves only a small amount in energy costs (ΔS_3).

The optimal economic thickness is arrived at when the last dollar invested for insulation results in one additional dollar in energy cost savings. This condition is met at thickness L_2, where ΔS_2 and ΔC_2 resulting from the addition of ΔL are equal, ($m_c = m_s$).

Life Cycle Costing

A popular concept of accounting for costs is the "life cycle" cost approach wherein the stream of outlays is added for the expected life of the project and the total cost over that life is compared with other alternatives. The cost analysis of this manual can be used in this mode as shown below.

The insulation project considers the cost of thermal losses and the cost of insulation together to solve for least-cost insulation thickness. The thermal losses are the product of the heat flow and value of that heat through the insulation. The cost of insulation considers the material, labor, and maintenance for the installation.

The thermal losses are given as heat flow per unit of insulation; the unit is either a linear foot of pipe or a square foot of flat surface. The heat flow through the insulation is assumed to be constant over the life of the installation. The cost of the heat is taken from the price of heat or fuel, conversion efficieny of fuel, heating value of fuel, and the fact that prices of heat or fuel are expected to change over the years of the project life.

In life cycle costing the procedure would be to take each year of life and escalate the cost of the heat for each year and then sum those costs. In this manual a multiplying factor, A, is provided, which sums all of the annual compounding multipliers and divides by the number of years of life to give an average multiplier for the term of the project. This saves calculating the stream of costs, adding then up, and dividing by the number of entries to get an average. Thus, the manual does the life cycle cost analysis for the cost of heat.

The cost of installed insulation must also be reduced to an average annual cost in order to compare it with the heat cost. Since the cost of installed insulation is an initial lump sum, that cost is multiplied by a factor, B_3, which gives the annual equal payment that will just return the capital and interest at the end of the project life. This uniform payment principle is also used in the life cycle costing for plant capital expense. For the sake of simplicity, the insulation maintenance costs are estimated to be 10 percent of the annual installed insulation cost for each year and added in as such.

It is evident then that the manual provides a least cost insulation thickness based upon the principles of life cycle costing (See Figure 2-1). To achieve this, the compounding interest rates for the factor A (to adjust heat cost) and for the factor B_3 (to amortize the capital investment of the insulation cost) must be equal to the rates used in the life cycle analysis. (B_2, used to amortize the cost of a heat-producing facility, is calculated using an interest rate, i_2, and term of the facility, n_2, both of which may be different from the rate and term used to calculate B_3).

Present Value (PV) Analysis

For the sophisticated cost analyst who prefers to use present value as the proper yardstick for evaluating projects, the method of this manual does just that, although it may not give the first impression of doing so.

Present value brings future cash flows back to the present by discounting at desired rates. The solution provided with this manual gives an identical answer by forcing the insulation thickness chosen to provide the least expensive solution for the sum of heat costs raised to inflated levels and the installed insulation costs raised to average cash levels using the cash of the future. To match a true present value analysis it is only required that the B_3 multiplication factor use the life term in years that the PV analysis uses and that the i_3 of the B_3 factor match the discount rate of the PV solution. The discount rate can be that for mere inflation or for an opportunity cost of money or for a required return on investment. The last increment of the calculated economic thickness will return that chosen rate. (See Chapters 3 and 5 for the calculation and use of discount factors.)

CHAPTER 3
COST OF ENERGY

This chapter solves for the cost of energy in a system. The solution for the cost of energy is ultimately combined with the cost of insulation in calculating the economic thickness.

Contents

1. Glossary
2. Choice of Insulation and Facility Life Terms
3. Choice of Cost of Money Terms
4. Cost of Heat Worksheet
5. Cost of Refrigeration Worksheet
6. Nomographs & Tables

Glossary

A   multiplier for average annual cost = $\dfrac{(1+i_1)^{n_1}-1}{i_1 \, n_1}$

AC_h annual cost of heat, $/10^6 Btu

B   annual amortization factor for capital investments = $\dfrac{i_2 \, (1+i_2)^{n_2}}{(1+i_2)^{n_2}-1}$

Btu   British thermal unit

C_h first year cost of heat, $/10^6$ Btu

C_k annual capital investment cost of heat, $/10^6$ Btu (This figure accounts for that portion of the total cost of heat which is attributed to the capital equipment needed to produce the heat.)

C_{kr} annual capital investment cost of refrigeration, $/10^6$ Btu

C_w cost of evaporated (make-up) water, $/10^6$ Btu

COP   coefficient of performance-refrigeration systems

E   conversion efficiency-process fuel to heat, decimal units only

°F   degrees fahrenheit

ft^2 square feet

ft^3	cubic feet
gal	gallon
H_c	heating value of coal, Btu/lb
H_g	heating value of gas, Btu/ft^3
H_o	heating value of oil, Btu/lb
hr	hour = 60 minutes
i_1	annual rate of price escalation, decimal units only
i_2	annual cost of money to finance plant, decimal units only
in	inch
kWh	kilowatt hour, 1,000 watt hr (3,413 Btu)
M	project average annual cost of heat, $/10^6 Btu
M_r	project average annual cost of refrigeration, $/10^6 Btu
n_1	number of years over which the insulation project is being amortized
n_2	life of the facility. This term can be the facility service life, depreciation period, or period over which the facility is being financed.
P_c	price of coal, $/ton
P_e	price of electricity, $/kWh
P_f	capital cost of heat producing facility, millions of dollars
P_g	price of gas, $/1,000 ft^3
P_h	price of purchased heat, $/10^6 Btu
P_o	price of oil, $/gal
P_r	capital cost of refrigeration system, dollars x 10^3
P_w	price of water, $/gal

Q   average annual expected heat production, millions of 10^6 Btu/year (annual fuel use x conversion efficiency x heating value of fuel)

S   constant depending on the type of refrigeration system

T_r refrigeration load, tons

w_o specific weight of oil, lb/gal

Y   annual operating time, hours

$ dollars

Choice of Insulation and Facility Life Terms, n_1 and n_2

The method of calculating optimal economic thickness, as presented in this manual, arrives at that thickness that produces the minimum cost over the assigned insulation life. There are two terms (time periods) which must be chosen: the insulation project life (n_1) and the life of the facility (n_2). The insulation project life is that period over which the initial insulation investment is annualized, or prorated. This figure can be varied according to the circumstances involved. Remember, when choosing an insulation project life, that the last increment (usually ½ inch) of insulation in the economic thickness will just recover its cost in energy cost savings in the assigned period, at the assigned cost of money. For a utility, insulation will most likely be in productive service until the insulation material itself degrades to the point where replacement is required. In this type situation, the insulation project life (n_1) should be the entire service life of the material (usually 15 to 20 years).

A second circumstance is that in which the insulation must pay for itself in energy saved in a period less than its potential material service life. In this case, the chosen n_1 can be either the assigned project depreciation period or the desired payback period. The shorter the time period given for the insulation to pay for itself, the thinner the economic thickness solution. However, it should be remembered that total present value costs will not be minimized over a period longer than that chosen for n_1.

The insulation project life (n_1) is also the period over which the cost of fuel is averaged. This is naturally so, because the economic thickness method minimizes the total cost of energy and insulation over the insulation project life.

The life of the facility (n_2) is the period over which the initial plant cost is annualized, for the purpose of assigning a capital cost to the cost of energy. The entire period for which the plant will be in operation should be chosen for the n_2 value, although for accounting purposes the depreciation period for the plant may be justified. A shorter payback period is not a proper value for n_2 because this would, in effect, create an artificially high value of energy, resulting in an overexpenditure for insulation.

Choice of Cost of Money Terms i_1, i_2, and i_3

The annual fuel price increase percentage, i_1, is used to calculate the average cost of fuel over the insulation project life. This is necessary because most projections agree that fuel will increase in price at least through the short and midterms. Therefore, the energy that will be saved with insulation in the future will be more valuable than it is today, justifying additional insulation investment.

It is difficult to place an exact value on the fuel price increase rate, and the user may be forced to arbitrarily chose a number, such as the expected rate of inflation. Should the fuel supply be on a long-term contract basis or if the plant is built at the energy source (such as a utility at a coal field), the price increase will be minimal.

The annual cost of money to finance the plant, i_2, is that rate at which the initial plant cost is annualized over the life of the facility (n_2). (Again, this annualized capital cost is divided by the expected annual energy production to arrive at the capital cost fraction of the total cost of energy.) The rate selected for i_2 should be the real cost of financing, whether this be the rate of interest on the borrowed funds used to build the energy plant, or the rate which is usually described as the cost of money.

The cost of money for the insulation project, i_3, used in Chapter 5 is the desired annual rate of return on the last increment of insulation in the recommended economic thickness. Rate of return may also be called opportunity cost. For example, if the economic thickness using an i_3 value of 20 percent is 3½ inches, that last ½ inch will annually return 20 percent of the initial investment required for that ½ inch. However, since insulation normally is available in minimum ½ inch increments, the last actual increment added will most likely return slightly more or less than the i_3 value. The higher the required rate of return, i_3, the thinner the economic insulation. If insulation is not required to meet a high return on investment criteria, the i_3 value should be a lower rate, such as the cost of borrowing or the cost of money.

For each of the rates i_1, i_2, and i_3, the user should consult his accounting department for the proper value.

Cost of Heat Worksheet

1. Multiplier for average annual heat cost, A, using Figure 3-1

   a. Enter Insulation Project Life, years n_1=_____years

   b. Enter Annual Fuel Price Increase i_1= 0.____

   c. Find Multiplier A   A=_____

2. First Year Cost of Heat, C_h, using Figure 3-2 (a, b, c or d)

   a. Enter Heating Value of Fuel   H=_____Btu/

   b. Enter Efficiency of Conversion, fuel to heat   E= 0.____

   c. Enter First Year Price of Fuel   P(o,g,c,e) =$_____/

   d. Find First Year Cost of Heat C_h=$_____/$10^6$Btu

3. Average Annual Heat Cost, using Figure 3-3

   a. Find Average Annual Cost of Heat for purchased steam and electric heat plants (no operating or maintenance costs) AC_h=$_____/$10^6$Btu

   b. Find Average Annual Heat Cost for coal, oil, and gas plants (10 percent operation and maintenance costs added) $(1.1)AC_h$=$_____/$10^6$Btu

4. Compound Interest Factor, $(1+i_2)^{n_2}$, using Figure 3-4

   a. Enter Life of Facility, years n_2=_____years

   b. Enter Annual Cost of Money to finance plant i_2= 0.____

   c. Find Compound Interest Factor $(1+i_2)^{n_2}$=_____

5. Annual Amortization Multiplier for Capital Investment, B, using Figure 3-5

Cost of Heat Worksheet, (Continued)

    a. Find B, using i_2 and $(1+i_2)^{n_2}$ from 4, above   B=_____

6. Annual Capital Cost of Heat, C_k, using Figure 3-6

    a. Enter Expected Average Annual Heat Production, Q   Q=___ millions of 10^6 Btu

    b. Enter Capital Investment of Heat Plant P_f=$___ million

    c. Find Annual Capital Cost of Heat C_k=$___$10^6$Btu

7. Find Project Cost of Heat, M,
$M = (1.1)AC_h + C_k$

                                 $(1.1)AC_h$ from step 3=$____/$10^6$Btu

                                 C_k from step 6=$____/$10^6$Btu

                                 M=$____/$10^6$Btu

Cost of Refrigeration Worksheet

1. Operating Cost

    a. Determine M, using procedure on Cost of Heat Worksheet $M=\$\underline{\quad}/10^6 Btu$

    b. Determine Coefficient of Performance using Figure 3-7 $COP=\underline{\quad}$

    c. Find Operating Cost of Refrigeration using Figure 3-8 $M/COP=\$\underline{\quad}/10^6 Btu$

2. Make-up Water Cost

    a. Find Average Annual Cost Multiplier using Figure 3-1 when

    i_1 = expected annual increase in water price $i_1 = 0.\underline{\quad}$
    n_1 = insulation project life $n_1 = \underline{\quad}$ years

    Find average Annual Cost Multiplier $A = \underline{\quad}$

    b. Find S using Table 3-1 $S = \underline{\quad}$

    c. Select First Year Price of Water $P_w = \$\underline{\quad}/gal$

    d. Find Average Make-up Water Cost, using Figure 3-9 $C_w = \$\underline{\quad}/10^6 Btu$

3. Refrigeration Equipment Capital Cost Assigned to Cost of Refrigeration, C_{kr}, using Figure 3-10

    a. Enter Annual Operating Time, hrs/yr $Y = \underline{\quad}$ hrs/yr

    b. Enter Load, tons $T_r = \underline{\quad}$ tons

    c. Find Annual Refrigeration, 1,000's ton-hr $YT_r = \underline{\quad}$ ton-hr

    d. Enter System Capital Cost, 1,000's $ $P_r = \$\underline{\quad}$ 1,000's

    e. Use B from Cost of Heat Worksheet Step 5 $B = \underline{\quad}$

    f. Find C_{kr}, Capital Cost of Refrigeration $C_{kr} = \$\underline{\quad}/10^6 Btu$

Cost of Refrigeration Worksheet, (Continued)

4. Find M_r

    M_r = M/COP + C_w + C_{kr}

    M/COP from Step 1 = $\$$____/10^6Btu
    C_w from Step 2 = $\$$____/10^6Btu
    C_{kr} from Step 3 = $\$$____/10^6Btu
    M_r = $\$$____/10^6Btu

Table 3-1. S Values For Refrigeration Systems

Electric motor driven chillers------------------------S=3.2

Absorption chillers----------------------------------S=6.2

Steam turbine driven chillers------------------------S=6.2

Gas and oil fired turbine driven chillers------------S=3.2

Gas and oil fired engine driven chillers-------------S=3.8

Note.—This value may be approximated by $S = 2 \frac{(COP + 1)}{COP}$ (See Appendix A).

Source: ASHRAE Guide and Data Book Systems 1970, Ch. 41, p. 639, American Society of Heating, Refrigeration and Air-Conditioning Engineers, Inc., New York, N.Y. 10017

Table 3-2. Typical Heating Values of Various Coals, Btu/lb x 10³

Anthracite Coal

Arkansas	13.7
Colorado	14.6
New Mexico	13.3
Pennsylvania	12-13.5
Virginia	12.0

Bituminous Coal

Alabama	12-14.0
Arkansas	13-14.0
Colorado	10-13.5
Illinois	10-12.0
Indiana	10-12.0
Iowa	9-11.0
Kansas	11-13.0
Kentucky	11-14.0
Maryland	13-14.0
Michigan	11.0
Missouri	10-13.0
Montana	11.0
New Mexico	10-12.0
N. Dakota	6- 7.5
Ohio	12-13.0
Oklahoma	12-14.0
Pennsylvania	12-14.0
Tennessee	13.0
Texas	7-12.0
Utah	13.0
Virginia	14.0
Washington	10-12.0
West Virginia	13-15.0
Wyoming	9-13.0

Table 3-3. **Available Energy in Saturated Steam**
(Used to calculate heat production when steam flow is known)

Absolute Steam Pressure (lb/in^2)	Steam Temperature (°F)	Latent Heat of Vaporization (Btu/lb)	Absolute Steam Pressure (lb/in^2)	Steam Temperature (°F)	Latent Heat of Vaporization (Btu/lb)
15	213	970	400	445	781
25	240	952	450	456	767
35	259	939	500	467	755
45	274	929	600	486	732
55	280	920	700	503	710
65	298	912	800	518	689
75	308	905	900	532	669
85	316	898	1000	545	649
95	324	892	1100	556	630
100	328	889	1200	567	612
120	341	878	1300	577	593
140	353	868	1400	587	575
160	364	859	1500	596	556
180	373	851	2000	636	463
200	382	843	2400	662	383
250	401	825	3000	695	218
300	417	809			
350	432	794			

Example.-given steam cost = \$4.25/1000 lb.
steam pressure = 250 psia (401°F)
Heat of Vaporization = 825 Btu/lb

find cost of steam/10^{6}Btu
$$= \frac{\$4.25}{825 \times 10^{-3}} + \$5.15/10^6 \text{ Btu}$$

REF: Keenan and Keyes, "Thermodynamic Properties of Steam," John Wiley & Sons, Inc., New York, N.Y., First Edition. 1936.

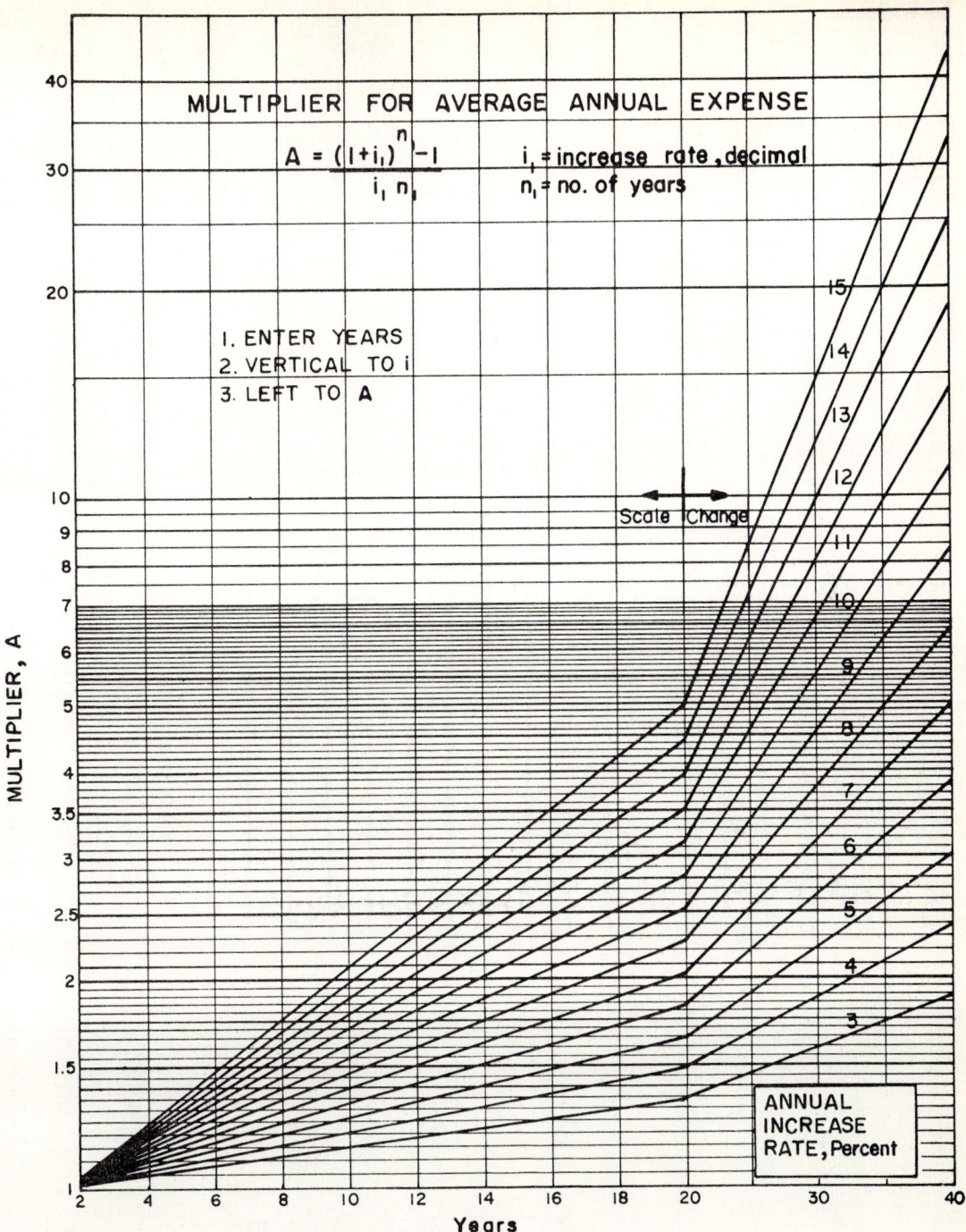

Figure 3-1. Multiplier to apply to present costs for determining the average annual costs when uniform cost increases occur in future years

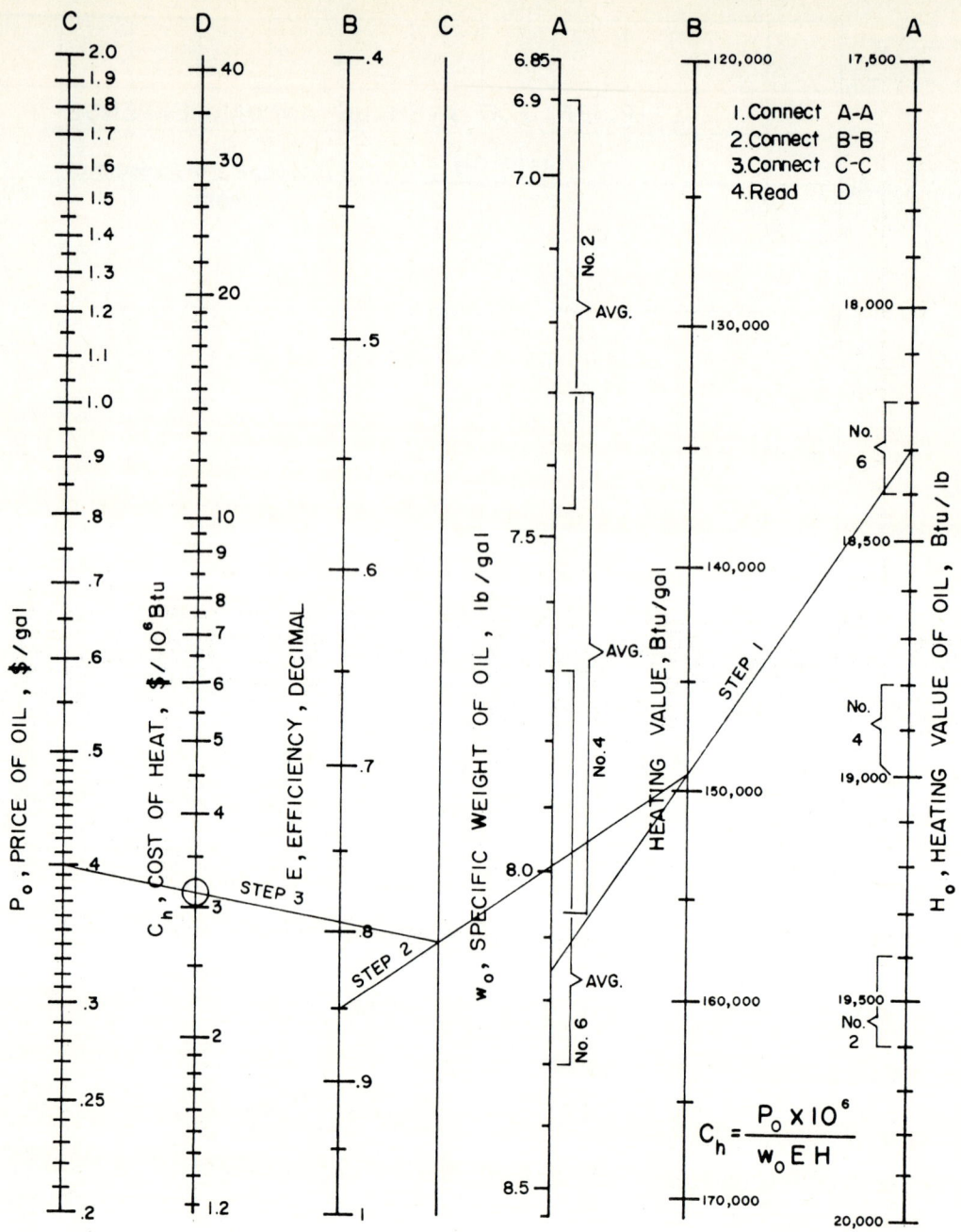

Figure 3-2a. The cost of heat, C_h, for fuel oil as the heat source

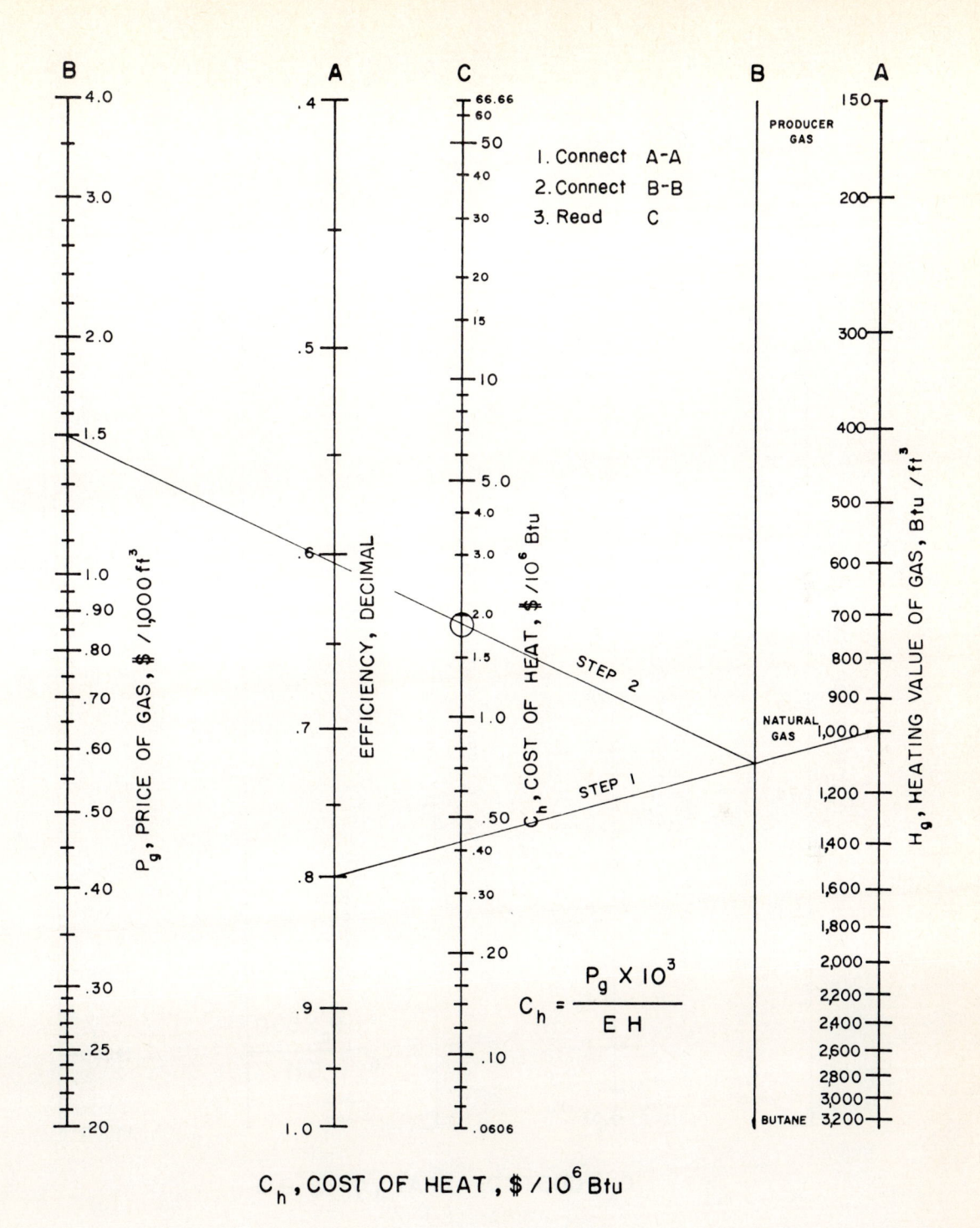

Figure 3-2b. The cost of heat, C_h, for gas as the heat source

$$C_h = \frac{P_g \times 10^3}{EH}$$

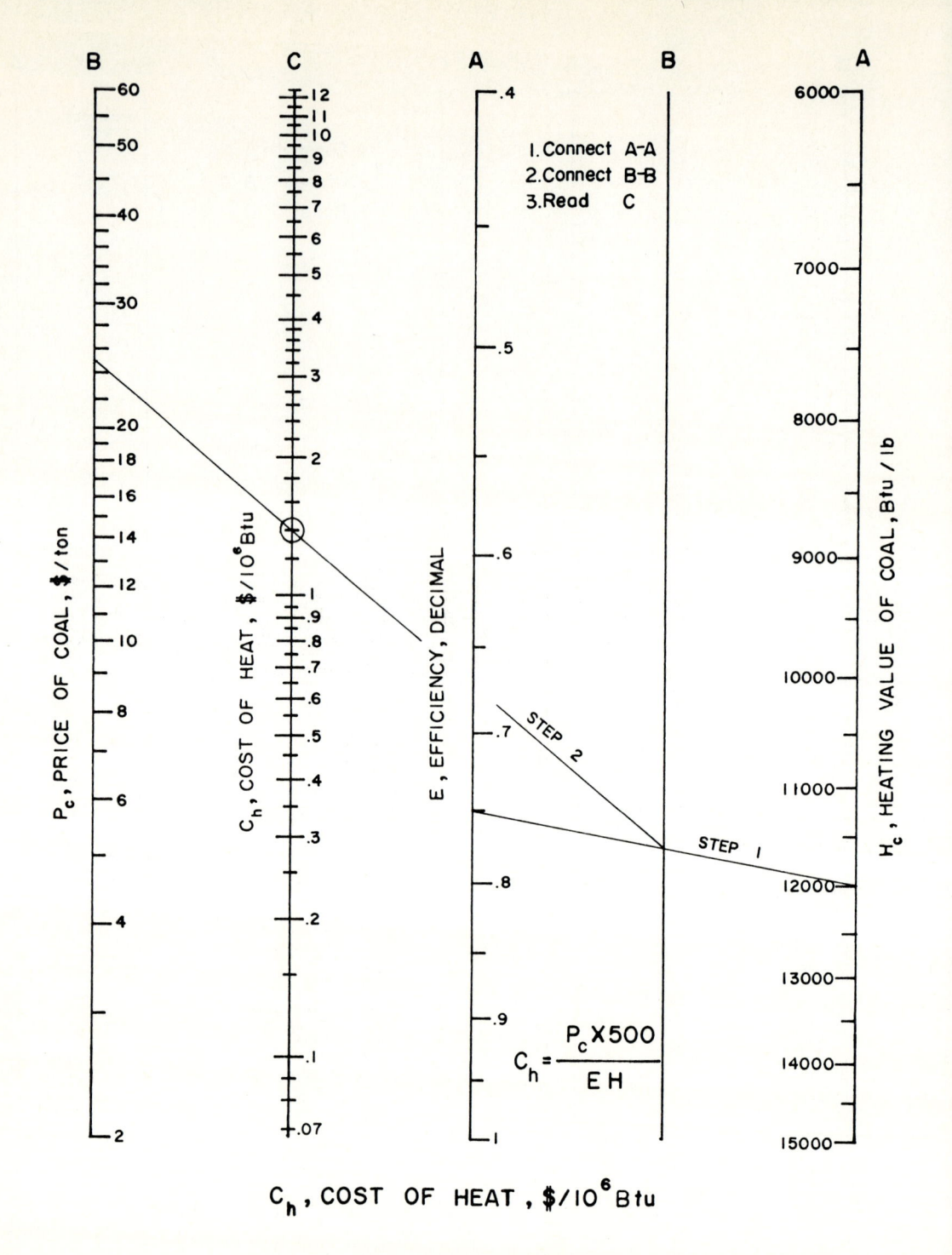

Figure 3-2c. The cost of heat, C_h, for coal as the heat source

$$C_h = \frac{P_c \times 500}{EH}$$

1. Connect A-A
2. Connect B-B
3. Read C

Figure 3-2d. The cost of heat, C_h, for electricity as the heat source

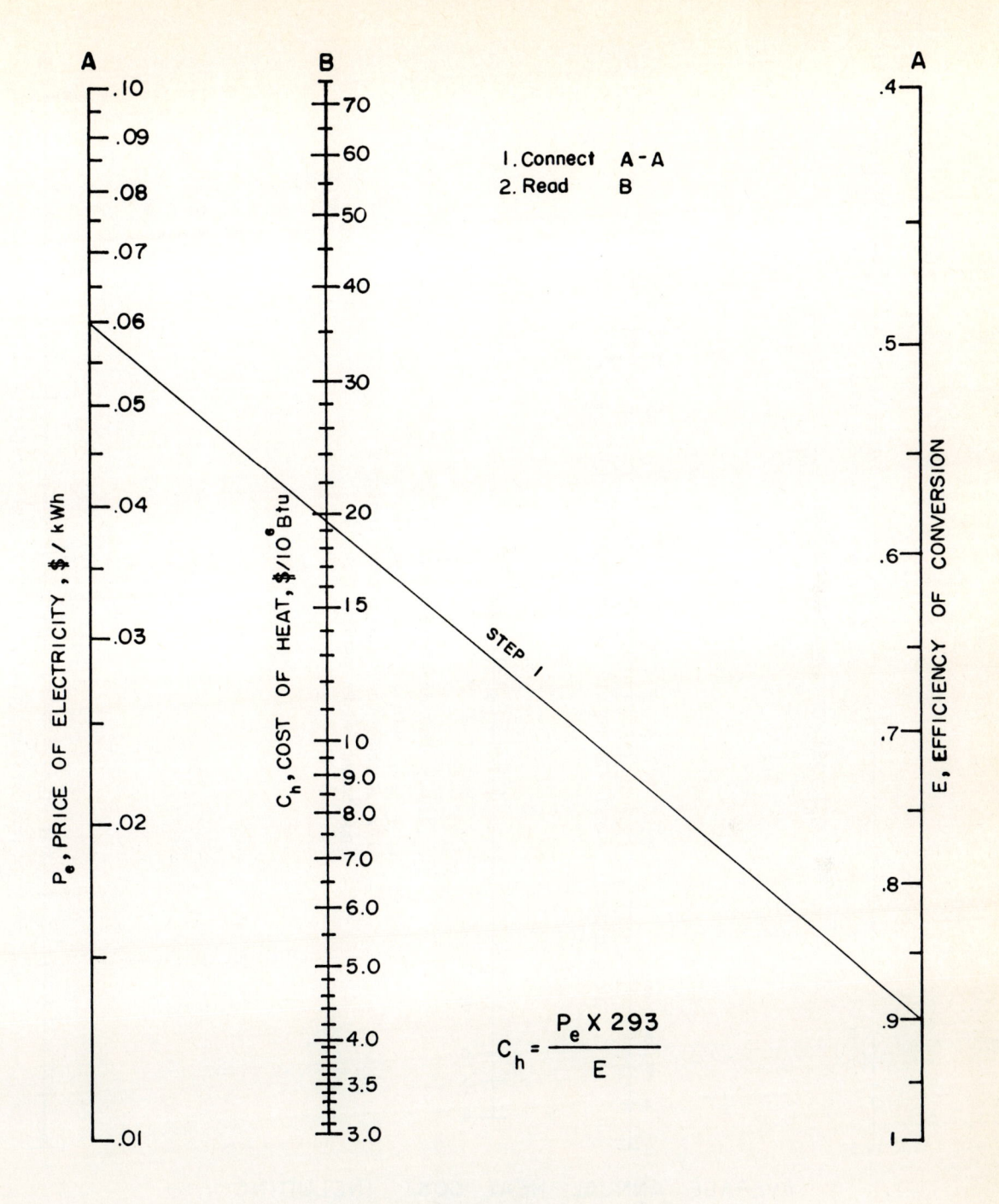

$$C_h = \frac{P_e \times 293}{E}$$

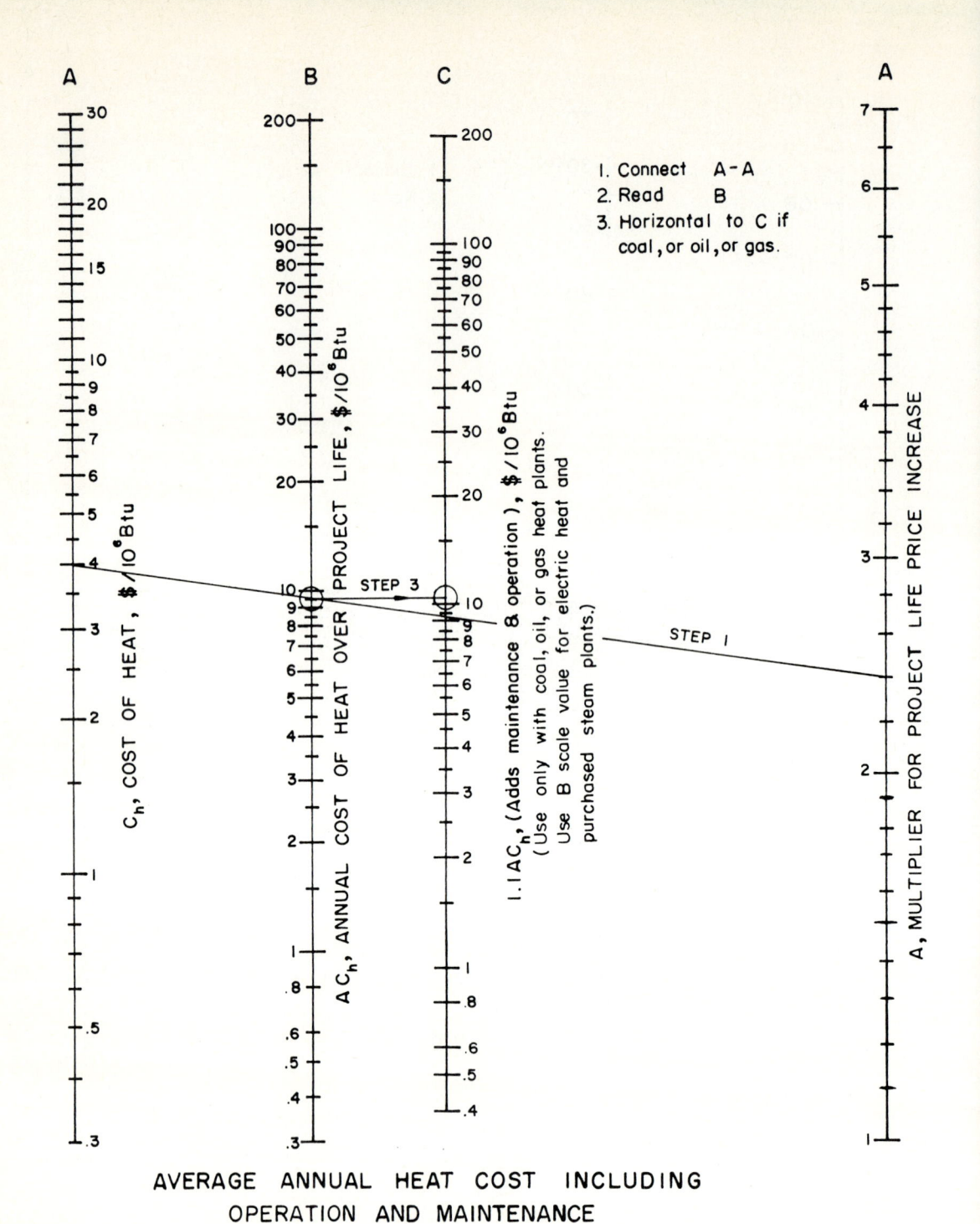

Figure 3-3. The average annual value of heat cost including operation and maintenance costs at the heat producing facility

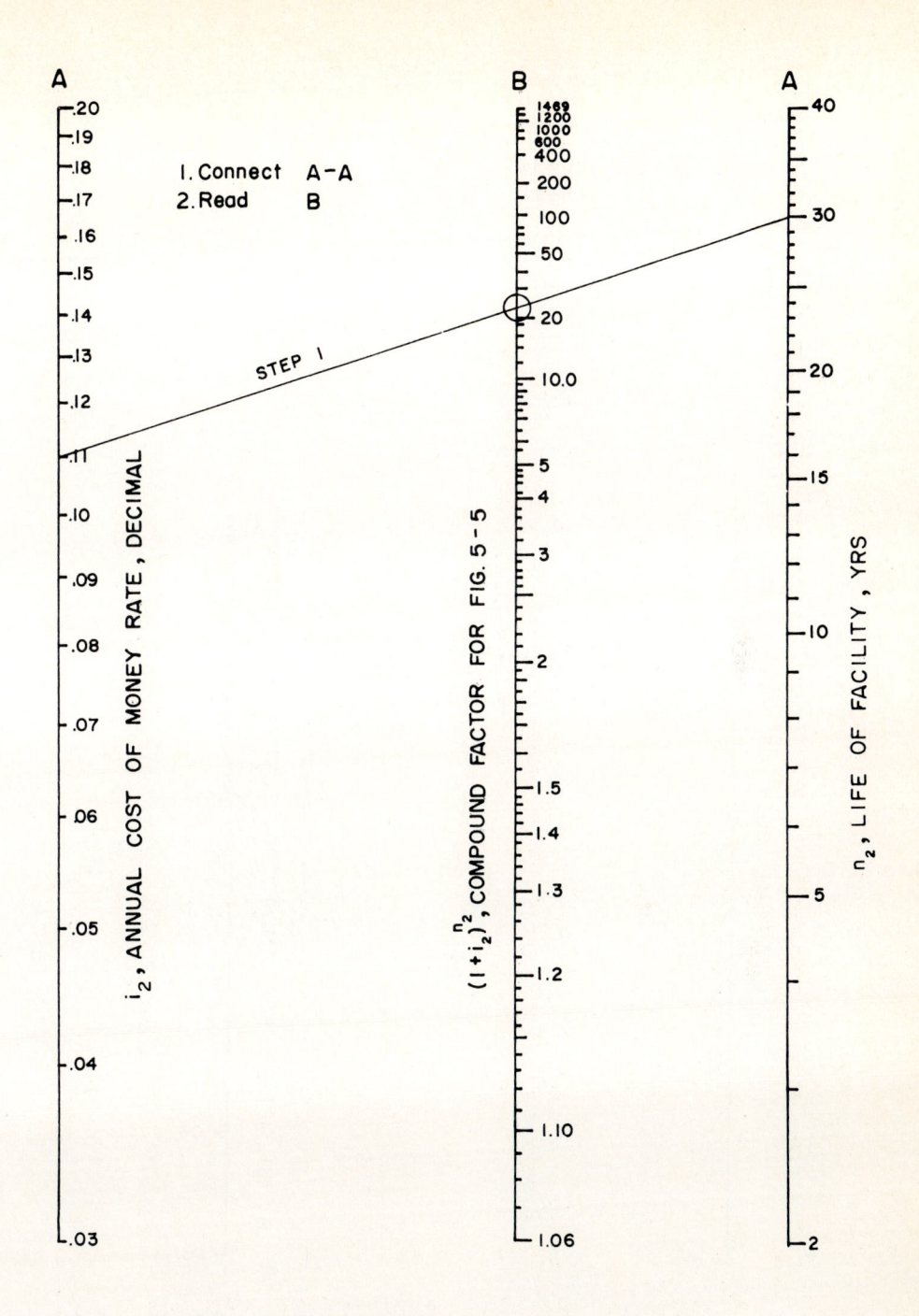

$(1+i_2)^{n_2}$, COMPOUND INTEREST FACTOR

Figure 3-4. Compound interest factor for use with Figure 3-5

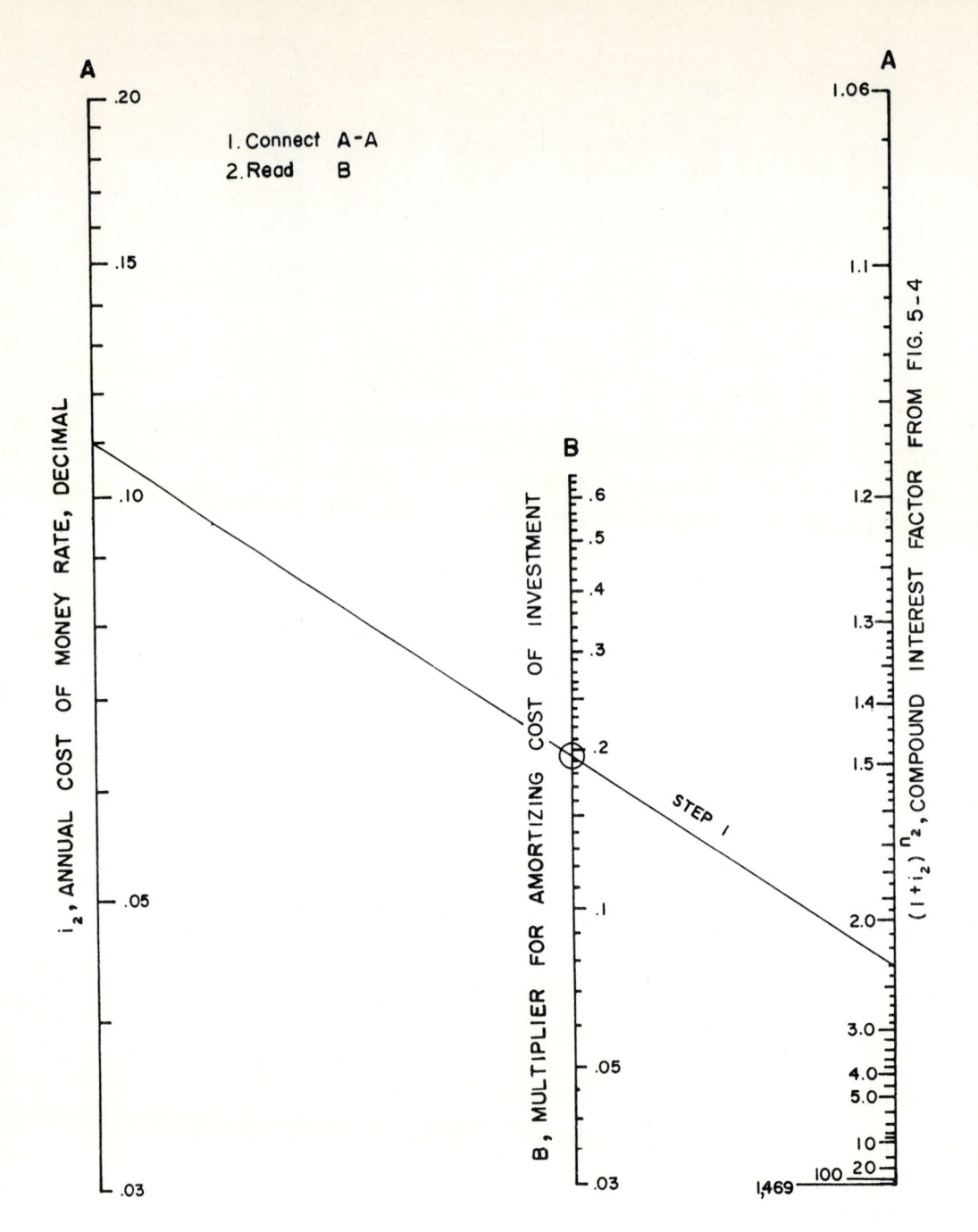

Note. The B multiplier $\dfrac{i_2 (1+i_2)^{n_2}}{(1+i_2)^{n_2} - 1}$ for amortizing an initial cost over a period of time with equal increments.

Figure 3-5. *The amortization multiplier, B*

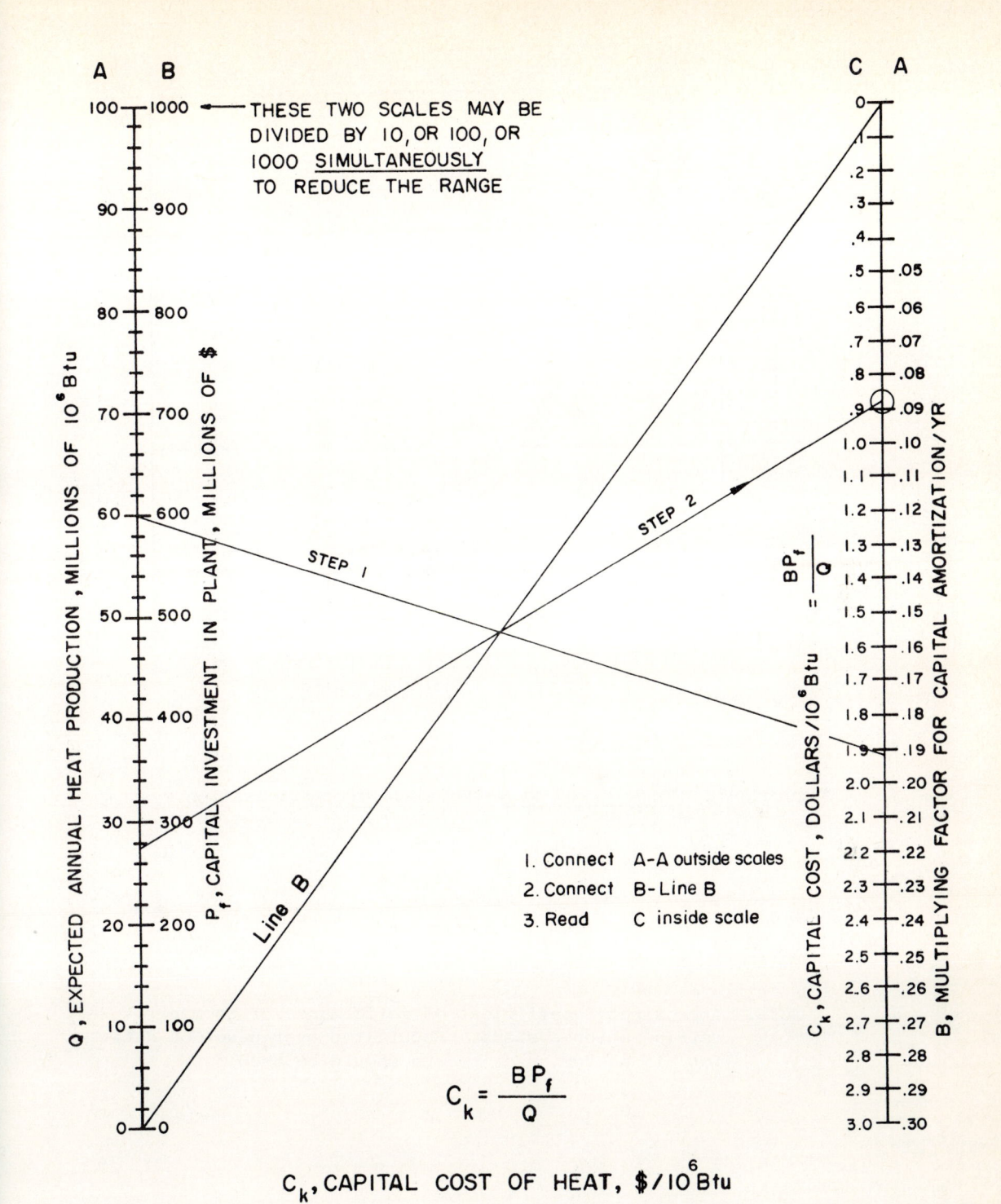

Figure 3-6. The distribution of heat production capital costs over the energy output on an annual basis

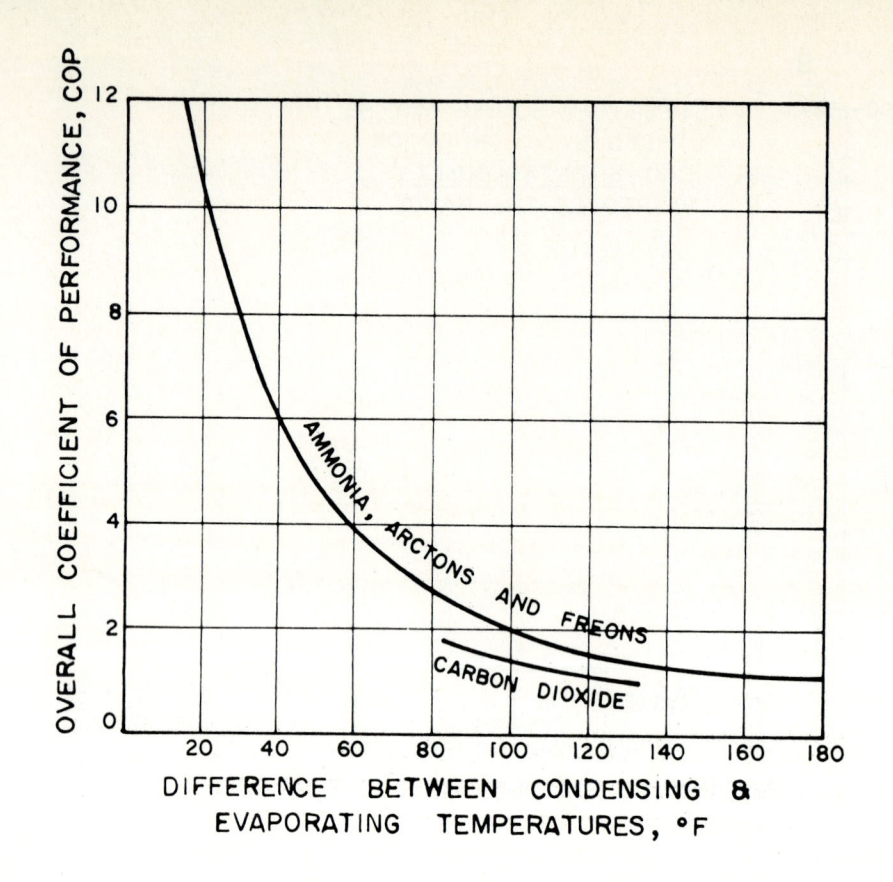

Figure 3-7. Coefficient of performance of refrigeration systems

Note. Approximate coefficient of performance of common refrigeration systems. Should the user have COP data for his system, user's data should be used.

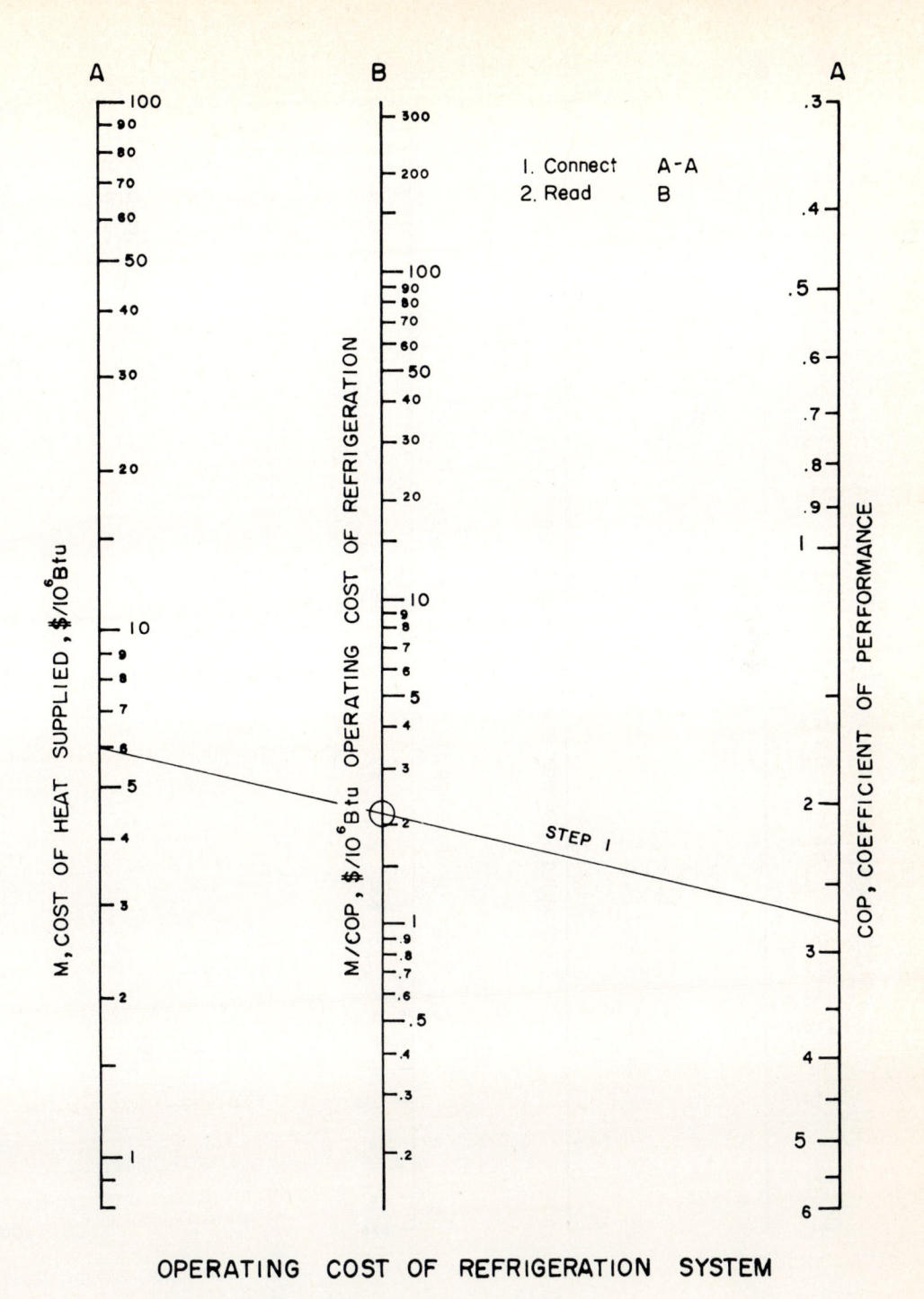

Figure 3-8. The operating cost of refrigeration systems

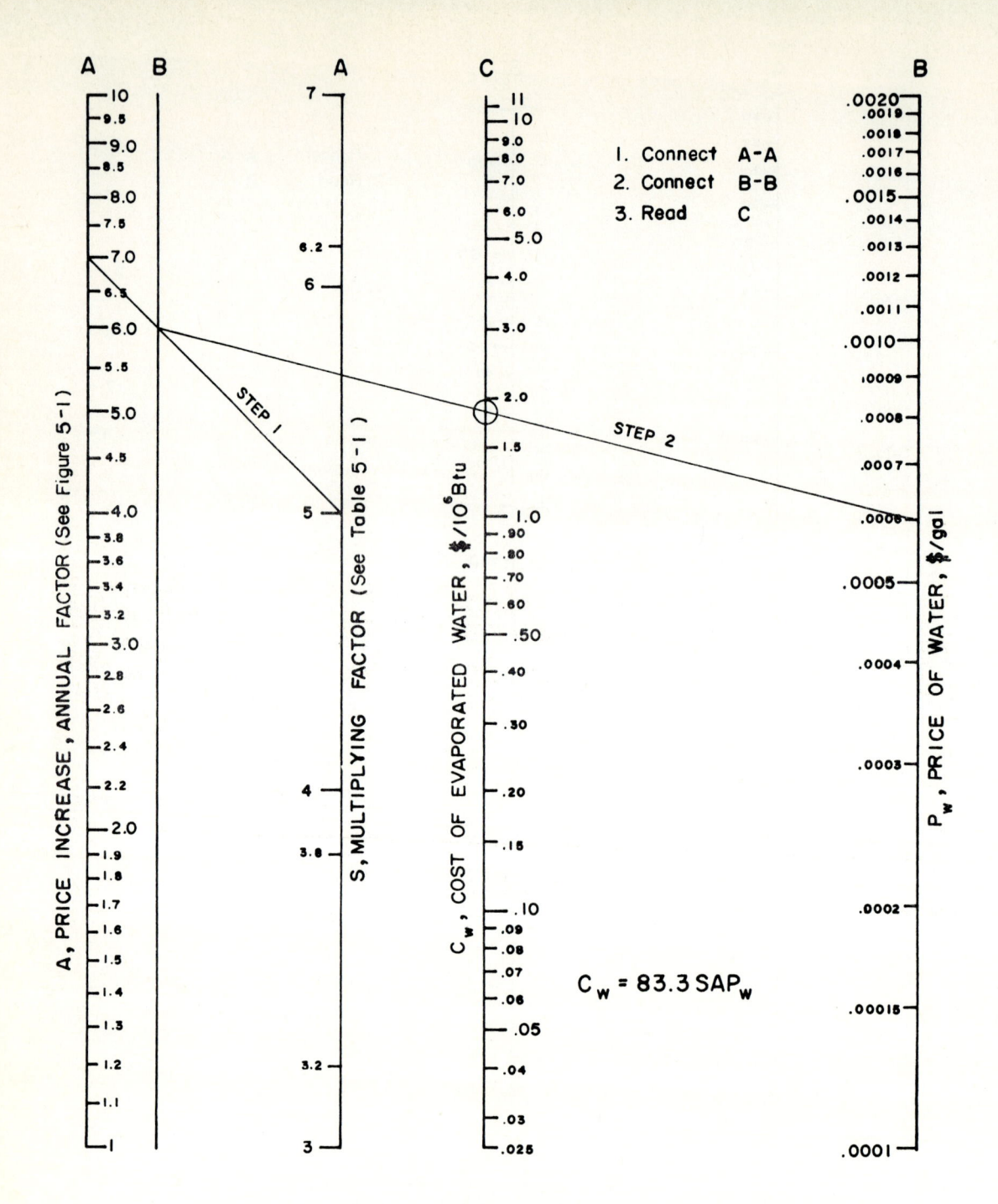

Figure 3-9. Average make-up water cost of water cooled refrigeration system

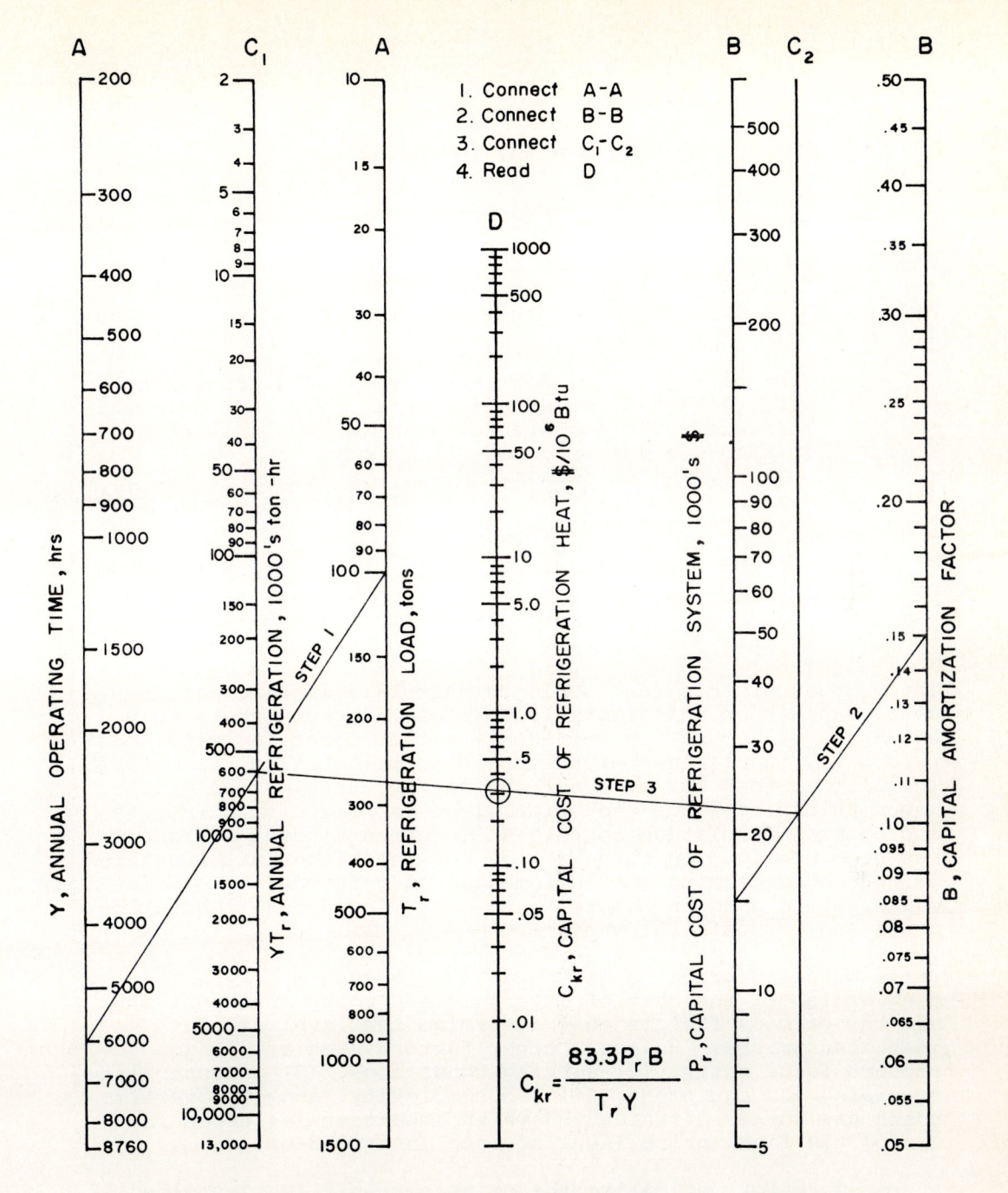

Figure 3-10. The annual average cost of capital investment in a refrigeration system

$$C_{kr} = \frac{83.3 P_r B}{T_r Y}$$

CHAPTER 4
THE COST OF INSULATION

Introduction

The cost of insulation is the most sensitive variable in the analysis for economic thickness (See Sensitivity Analysis, Appendix B). In order to perform this analysis it is necessary to

1. Know how the price for installed insulation varies with thickness for specific pipe sizes and equipment surfaces. The incremental cost factor, m_c, defined in this section provides this need.

2. Reduce the insulation investment into an annual expense, including maintenance. This is performed in Chapter 5.

3. Correct the price of installed insulation to include the increased cost of insulating valves, flanges and fititings. Piping complexity factors are provided in this section.

The purpose of this section is to calculate incremental insulation costs based upon unit installed prices. Because there is no known way of accurately representing the change in installed prices vs. insulation thickness with a mathematical formulation, installed prices for at least two single, double, and triple layer thicknesses for each pipe size are needed to calculate incremental insulation costs. An incremental cost worksheet has been provided in this section for recording unit installed prices and determing the incremental cost factor, m_c. This worksheet is used in Chapter 5 to determine annual insulation costs and economic thicknesses.

Unit installed price depends upon a multitude of variables. The cost of labor, the cost of materials and worker productivity are the primary factors that determine the level of unit installed prices. However, other factors such as (1) job support force costs, (2) working conditions, (3) accessibility of piping and equipment, (4) job complexity, and (5) overtime costs and worker efficiency have an important influence in fixing the final price level of each installed unit.

A procedure for estimating the price per unit for installed, frequently used industrial insulations and jackets is provided in this chapter. Its purpose is not to fix a budget price for insulating, but to typify regional unit installed prices so thicknesses can be specified and plant design layout work can commence. It is recommended that where possible accurate determinations of installed prices per unit typical of when the insulation is to be applied and incorporating all of the above factors be obtained by consulting a qualified insulation contractor.

Unit installed prices are an average estimate of pipe insulating costs. These prices do not include the cost effect of insulating valves, flanges, and other pipe fittings. If these piping components are to be insulated, the economic thickness for a piping section must also be based upon their insulating costs. Since it is impractical to evaluate the cost effect of fittings for every specific piping section in a plant, piping complexity factors for complex (in process piping), average, and simple (out process piping) have been provided in Table 4-1. (The data and method used to derive these factors are shown in Appendix F.) Interpolation is required in choosing the factors to be used in this analysis.

Glossary

F_c	composite labor factor
F_r	regional labor productivity factor
F_t	material type factor
L	insulation thickness, inches
LP	estimated labor price, \$/lin-ft or \$/ft^2
LR	local labor rate, \$/man-hr
m_c	incremental insulation cost, \$/lin-ft/in or \$/ft^2/in
MP	estimated material price, \$/lin-ft or \$/ft^2
P	price of installed insulation, \$/lin-ft or \$/ft^2
PC	piping complexity factor
R	material price ratio
WP	base worker productivity

Determining Incremental Insulation Costs

The incremental insulation cost, m_c, is determined directly from unit price of installed insulation. This factor represents the average dollar increase in price for an inch of insulation. Since prices for installed insulation increase substantially when insulation is applied in layers, an incremental cost factor must be determined for each layered insulation application. For each pipe size the following procedure must be performed to determine the appropriate single, double, and triple layer incremental cost factors.

1. Choose the applicable pipe size.

2. Obtain the installed price (P_1) for a thickness at the lower end of the single layer thickness range (L_1). Enter both (L_1) and (P_1) on the Insulation Cost Worksheet.

3. Obtain the installed price (P_2) for a thickness at the upper end of the single layer thickness range (L_2). Enter both (L_2) and (P_2) on the Worksheet. (Consult insulation contractor for steps 2 and 3.)

4. Choose the appropriate piping complexity factor, PC, from Table 4-1.

5. Calculate the single layer incremental insulation cost factor, m_c, and enter on the Worksheet.

$$m_c = PC \left(\frac{P_2 - P_1}{L_2 - L_1} \right)$$

6. Repeat (2) thru (5) for both the double and triple layer applications of insulation.

INSULATION COST WORKSHEET   DATE _____

PLANT _____
LOCATION _____   L   Insulation thickness, inches
APPLICATION _____   P   Installed price
SPECIFICATION _____   Piping - $/Linear foot
Insulation _____   Vessels - $/ft²
Jacket and Finish _____   (obtain from contractor or by using Estimator)
                          m_c Incremental cost
                              Piping - $/lin. ft. per inch
                              Vessels - $/ft² per inch

Pipe Size (inches)	Single Layer Prices				Double Layer Prices					Triple Layer Prices				
	L_1	P_1	L_2	P_2	m_{c1}	L_1	P_1	L_2	P_2	m_{c2}	L_1	P_1	L_2	P_2	m_{c3}
½															
3/4															
1															
1-1/2															
2															
2-1/2															
3															
4															
5															
6															
8															
10															
12															
14															
16															
18															
20															
24															
30															
36															
Flat surfaces and vessels															

Note.--1. L_1 = thickness of lower end of layer range; L_2 = thickness of upper end of layer range ;
          P_1 = installed price for L_1; P_2 = installed price for L_2 .
      2. m_c = PC(P_2-P_1/L_2-L_1).
      3. PC = Piping Complexity Factor (Table 4-1)

Table 4-1. Piping Complexity Factors, PC

Pipe Size, inches	Piping Complexities* Complex	Average	Simple
1/2 to 1-1/2	1.60	1.30	1.20
2 to 3	1.35	1.20	1.15
4 to 6	1.25	1.15	1.10
8 to 12	1.28	1.15	1.12
14 and above	1.30	1.20	1.15

*The above complexities are defined as

    Complex--30-40 fittings every 100 feet of pipe;
    Average--15-20 fittings every 100 feet of pipe;
    Simple --10 fittings every 100 feet of pipe.

Note. The above factors apply only to welded piping with flanged valves and line flanges. For welded piping with no flanges use an average PC factor of 1.05. These factors are not to be used if valves and flanges are not to be insulated.

    Fittings are denoted as

  flange pair--1 fitting;
  valve flanges--2 fittings;
  valve body--1 fitting;
  elbows, tees, reducers, etc.--1 fitting.

Estimating Price per Unit of Installed Insulation

There are two basic cost elements that determine installed insulation prices:

1. Material prices.
2. Labor for installing the material.

Material Prices

Material prices are directly related to the volume and cost of insulation, jacketing, securement, finishing, and structural support material required to meet the insulation specification. Also included are the costs to the contractor for storage, shipment, and handling of these materials. Although material prices vary with geographic area, the ratio of material price for a given insulation thickness and pipe size to that for another thickness and pipe size does not change appreciably from contractor to contractor.

Material price ratios have been provided for estimating unit material prices (see Tables 4-2 through 4-5). Included are tables for each of the following insulations finished with a 0.016 aluminum jacket:

1. Calcium silicate,
2. Preformed mineral wool,
3. Preformed fiber glass,
4. Cellular glass.

A factor of 1 is assigned to the material price for 2 inches of insulation for a 2-inch pipe. Material prices for other pipe sizes and thicknesses can be estimated from the factor provided in each table, once the unit material price for 2" x 2" (2 inches insulation on a 2-inch pipe) is obtained from a local contractor.

Material price ratios for the application of insulation on equipment and vessels (flat surfaces) are also provided in Tables 4-2 through 4-5. A factor of 1 has also been assigned to the material prices for 2 inches of insulation applied on flat surfaces. Unit material prices for other flat surfaces thicknesses can be estimated employing these factors, once the material prices for 2 inches of insulation on flat surface are obtained from a local contractor.

It is suggested that where possible several material prices be obtained from contractors to test the accuracy of these tables. If actual material prices deviate from the prices estimated with these tables, an appropriate multiplication factor may be devised from averaging the ratio of actual price to estimated price for a few pipe sizes and thickness combinations.

Labor for Installation

The labor cost for installing insulation varies considerably nationally and is influenced by

1. Local labor rate ($/man-hour);
2. Worker productivity (man-hours per linear foot or square foot of insulation);
3. Job size support force costs.

The local labor rate has a significant effect on determining the level of labor prices for various insulation thicknesses and pipe sizes. Labor rates are dictated by such factors as local labor availability, whether or not union labor is employed, and whether overtime work is involved. The labor rate includes the base labor wage, fringe benefits, per diem and travel expenses, and may also include elements of the job size support force cost.

Worker productivity (i.e., man hours required to insulate a segment of pipe) is the most variable element comprising unit labor price. The man hours required to insulate increase with large pipe sizes, inaccessibility of piping, and equipment and job complexity. Other factors which are known to influence worker productivity are

1. Type of material used,
2. Working conditions,
3. Local worker productivity,
4. Overtime loss of worker efficiency (productivity usually decreases as length of work week increases).

The job size support force cost is the third element that fixes final unit labor costs. Job size support force costs increase with the magnitude of the job. Although each contractor may treat this cost differently some of the major factors it includes are

1. Preparation costs (scaffolding, etc.),
2. Clean-up and tear-down costs,
3. Supervision costs,
4. General overhead.

Average worker productivity factors for insulating piping and flat surfaces are presented in Table 4-8. These factors, expressed in man hours per 100 linear feet of pipe, or 100 ft^2 of flat surface, were derived from a large sampling of insulating productivity across the United States and eastern Canada. The data used to construct this table are based on the working conditions noted below the table. Correction factors for material and regional productivity differences are found in Tables 4-6 and 4-7, respectively.

In utilizing Tables 4-6 through 4-8 to calculate unit labor costs, the labor rate ($/man-hr) must be obtained from a local contractor. The rate obtained should include base wage, fringes, per diem, travel expenses, profit.

Installed Cost Estimating Procedure

1. Select the insulation and jacket specification. (Estimating data have been provided for commonly occurring specifications, Tables 4-2 through 4-5.)

2. Obtain the following from a local insulation contractor:

   Insulation and jacket material prices for

   2" on 2" pipe $_____/lin-ft

   2" on a flat surface $_____/ft^2

   Local labor rate, LR

   (Total cost per hour that the
   contractor will charge)   LR=$_____/man-hr

Estimating Material Prices

3. Select the appropriate material price ratio table for the specification (Tables 4-2 through 4-5).

4. For each pipe size and insulation thickness shown on the top of the Insulation Cost Estimator Worksheet, multiply the material price for 2" on 2" pipe by the appropriate R Factor obtained from the material price ratio table. Enter the estimated material prices on the Cost Estimator Worksheet.

5. The same procedure applies when estimating material prices for insulating flat surfaces. Material prices for flat surface thickness is obtained by multiplying the appropriate R Factor by the material price for 2" on a flat surface.

Estimating Labor Prices

6. Calculate the Composite Labor Factor, F_c, for piping.

   From Table 4-6, enter the material type factor, F_t.

   F_t=_____

   From Table 4-7, enter the Regional Productivity Factor, F_r.

   F_r=_____

Enter local labor rate
(Obtained in Step 2)   LR= $_____ /man-hr

$$F_c = \frac{1.2 \times F_t \times F_r \times LR}{100}$$ F_c=_____

(Assuming a support force factor of 1.2)

7. For each pipe size and insulation thickness shown in the top of the Insulation Cost Estimator Worksheet, calculate labor prices (LP) by multiplying (F_c) by the appropriate worker productivity factor (WP) found in Table 4-8. Enter each labor price on the Worksheet.

$$F_c \times WP = LP$$

8. Repeat (6) and (7) for estimating flat surface labor prices.

9. Add material and labor price estimates to find estimated unit installed insulation price, P.

$$P = MP + LP$$

10. Enter P values on the Insulation Cost Worksheet.

Insulation Cost Estimator Worksheet

Material and Labor Prices*, MP and LP

Pipe Size (inches)	Single Layer 2 inch MP / LP	3 inch MP / LP	Double Layer 4 inch MP / LP	6 inch MP / LP	Triple Layer 7 inch MP / LP	9 inch MP / LP
1/2						
3/4						
1						
1-1/2						
2						
2-1/2						
3						
4						
5						
6						
8						
10						
12						
14						
16						
18						
20						
24						
30						
36						
Flat Surface						

*Piping Prices—$/lin-ft
Flat Surface Prices—$/ft²

Table 4-2. Unit Selling Price Ratios

(Specification: Calcium Silicate)

Pipe Size (inches)	Single Layer 1	Single Layer 1-1/2	Single Layer 2	Single Layer 2-1/2	Insulation Thickness, Inches 3	4	Double Layer 3	4	5	6	Triple Layer 6	7	8	9	10
1/2	0.44	0.54	0.78	0.93	1.10	1.98	1.24	1.98	2.70	3.45	–	–	–	–	–
3/4	0.45	0.56	0.81	0.97	1.18	2.00	1.32	2.03	2.75	3.50	–	–	–	–	–
1	0.47	0.59	0.85	1.00	1.25	2.10	1.46	2.14	2.90	3.62	–	–	–	–	–
1-1/2	0.49	0.64	0.90	1.08	1.34	2.20	1.51	2.28	3.10	3.85	–	–	–	–	–
2	0.54	0.70	1.0*	1.15	1.44	2.25	1.56	2.35	2.20	4.00	–	–	–	–	–
2-1/2	0.60	0.78	1.09	1.26	1.55	2.55	1.64	2.45	3.30	4.20	–	–	–	–	–
3	0.67	0.86	1.18	1.37	1.71	2.75	1.74	2.60	3.50	4.35	–	–	–	–	–
4	0.75	0.96	1.31	1.60	1.90	2.90	1.93	2.88	3.80	4.85	5.10	6.25	7.40	8.6	9.70
5	0.83	1.05	1.46	1.73	2.10	3.10	2.15	3.20	4.20	5.30	5.50	6.70	8.00	9.2	10.40
6	0.92	1.15	1.66	1.90	2.30	3.30	2.39	3.45	4.50	5.62	5.80	7.20	8.50	9.9	11.20
8	–	1.35	1.95	2.40	2.83	3.80	2.89	4.00	5.20	6.80	6.70	8.20	9.70	11.2	12.60
10	–	1.70	2.35	2.85	3.30	4.35	3.23	4.60	6.00	7.40	7.70	9.50	11.20	12.0	14.60
12	–	2.10	2.62	3.15	3.67	4.80	3.41	5.20	6.70	8.40	8.50	10.40	12.30	14.2	16.10
14	–	2.43	2.96	3.50	4.05	5.25	3.74	5.55	7.30	9.10	9.20	11.20	13.20	15.2	17.20
16	–	2.80	3.20	3.90	4.55	5.90	4.35	6.20	8.00	9.90	9.70	12.00	14.20	16.5	18.70
18	–	3.15	3.60	4.25	5.0	6.20	4.65	6.60	8.60	10.70	10.50	12.90	15.30	17.7	20.00
20	–	3.50	3.90	4.60	5.35	6.80	4.72	7.20	9.40	11.75	11.50	14.00	16.60	19.0	21.50
24	–	4.00	4.40	5.20	6.00	7.75	5.70	8.30	10.80	13.40	13.20	16.20	19.20	22.0	24.50
*30	–	–	5.70	6.75	8.00	10.00	7.95	10.50	13.20	16.00	15.50	18.30	21.10	24.0	26.80
*36	–	–	6.70	8.00	9.60	12.40	9.90	13.00	16.00	19.00	18.00	21.50	24.60	28.0	31.50
Flat Surface	.70	.84	1.0*	1.17	1.35	1.70	1.37	1.75	2.10	2.50	2.55	2.95	3.35	3.75	4.15

*Factors for piping based upon material price for 2 inches on 2 inch pipe, factors for flat surfaces based upon material price for 2 inches on flat surface.

Table 4-3. Unit Material Selling Price Ratios

(Specification: Preformed Mineral Wool)

| Pipe Size (inches) | Insulation Thickness, Inches ||||||||||||
| | Single Layer ||| Double Layer |||| Triple Layer ||||
	1	1-1/2	2	2-1/2	3	4	3	4	5	6	6	7	8	9	10
1/2	0.34	0.53	0.76	0.97	1.16	1.82	1.22	1.85	2.60	3.34	-	-	-	-	-
3/4	0.36	0.56	0.90	1.00	1.27	1.87	1.38	1.96	2.68	3.53	-	-	-	-	-
1	0.42	0.59	0.85	1.07	1.33	2.96	1.48	2.05	2.81	3.67	-	-	-	-	-
1-1/2	0.44	0.67	0.94	1.16	1.45	2.04	1.59	2.15	2.96	3.83	-	-	-	-	-
2	0.50	0.72	1.0*	1.22	1.55	2.11	1.67	2.24	3.11	3.67	-	-	-	-	-
2-1/2	0.55	0.79	1.06	1.36	1.69	2.20	1.78	2.36	3.31	4.24	-	-	-	-	-
3	0.61	0.86	1.15	1.49	1.89	2.33	1.90	2.51	3.52	4.48	-	-	-	-	-
4	0.77	1.01	1.29	1.71	2.22	2.71	2.15	2.79	3.95	4.71	4.87	5.60	6.38	7.15	7.92
5	0.89	1.14	1.44	1.92	2.46	3.05	2.34	3.08	4.29	5.01	5.03	5.81	6.60	7.40	8.20
6	1.02	1.24	1.56	2.08	2.76	3.23	2.77	3.33	4.61	5.37	5.41	6.21	7.02	7.80	8.57
8	1.25	1.39	1.81	2.31	3.07	3.69	3.01	3.79	5.19	6.06	6.30	7.23	8.11	8.99	9.86
10	-	1.63	2.13	2.72	3.48	4.26	3.45	4.42	5.77	6.89	7.10	8.20	9.32	10.42	11.54
12	-	1.80	2.43	3.09	3.83	4.69	3.91	4.97	6.32	7.76	7.43	9.30	10.72	12.11	13.51
14	-	2.13	2.66	3.37	4.17	5.21	4.18	5.40	6.85	8.40	9.52	10.01	11.52	13.05	14.55
16	-	2.33	2.91	3.60	4.55	5.42	4.57	5.86	7.38	9.10	9.10	10.79	12.46	14.15	15.89
18	-	2.54	3.14	3.95	4.92	5.98	4.96	6.26	7.97	9.58	9.62	11.35	13.08	14.78	16.52
20	-	2.86	3.51	4.32	5.43	6.35	5.52	6.71	8.55	10.16	10.22	12.00	13.77	15.62	17.42
24	-	3.36	3.97	4.96	6.22	6.90	6.35	7.38	9.38	11.12	11.14	12.91	14.73	16.60	18.40
*30	-	4.13	4.72	5.77	7.31	7.63	7.62	8.48	10.65	12.72	12.75	14.68	16.75	18.74	21.70
*36	-	4.71	5.28	6.26	7.81	8.53	9.22	9.27	11.46	13.65	13.70	15.87	18.25	20.54	23.94
Flat Surface	.70	.86	1.0*	1.16	1.13	1.80	1.36	1.72	1.94	2.25	2.35	2.67	2.96	3.24	3.52

*Factors for piping based upon material price for 2 inches on 2 inch pipe, factors for flat surfaces based upon material price for 2 inches on flat surface.

Table 4-4. Unit Material Selling Price Ratios
(Specification: Preformed Fiber Glass)

Pipe Size (inches)	1	Single Layer 1-1/2	2	2-1/2	3	4	Double Layer 3	4	5	6	6	Triple Layer 7	8	9	10
1/2	0.32	0.54	0.79	0.90	1.06	1.98	1.33	2.00	2.70	3.40	—	—	—	—	—
3/4	0.34	0.56	0.82	0.95	1.09	2.03	1.35	2.05	2.75	3.50	—	—	—	—	—
1	0.37	0.59	0.86	1.01	1.16	2.11	1.38	2.14	2.80	3.65	—	—	—	—	—
1-1/2	0.43	0.66	0.94	1.08	1.24	2.20	1.43	2.25	3.00	3.75	—	—	—	—	—
2	0.47	0.72	1.0*	1.17	1.34	2.26	1.50	2.31	3.10	3.90	—	—	—	—	—
2-1/2	0.52	0.78	1.06	1.27	1.48	2.34	1.66	2.48	3.27	3.90	—	—	—	—	—
3	0.57	0.83	1.15	1.40	1.64	2.45	1.77	2.59	3.40	4.25	—	—	—	—	—
4	0.71	0.94	1.32	1.60	1.90	2.66	1.90	2.85	3.75	4.65	4.85	6.20	7.55	8.90	10.3
5	0.80	1.04	1.48	1.80	2.14	3.01	2.15	3.10	4.05	5.00	5.25	6.68	8.10	9.53	10.9
6	0.88	1.10	1.54	1.94	2.28	3.25	2.30	3.30	4.36	5.35	5.60	7.10	8.60	10.10	11.6
8	1.18	1.40	1.88	2.30	2.74	3.51	2.75	3.90	5.00	6.15	6.40	8.13	9.85	11.60	13.3
10	—	1.58	2.13	2.72	3.20	4.26	3.15	4.35	5.60	6.80	6.88	8.68	10.50	12.30	14.0
12	—	1.88	2.50	3.03	3.55	4.61	3.50	4.90	6.25	7.60	7.65	9.70	11.70	13.70	15.8
14	—	2.10	2.75	3.34	3.93	5.21	4.00	5.40	6.85	8.30	8.35	10.50	12.70	14.90	17.0
16	—	2.34	3.0	3.7	4.35	5.53	4.35	5.95	7.50	9.10	9.10	11.50	13.90	16.3	18.7
18	—	2.58	3.32	4.04	4.78	6.18	4.75	6.45	8.20	9.90	9.90	12.50	15.00	17.60	20.0
20	—	2.83	3.64	4.38	5.10	6.62	5.15	6.95	8.75	10.60	10.28	13.00	15.80	18.60	21.4
24	—	3.35	4.10	4.82	5.62	7.17	5.85	7.90	9.98	12.00	12.00	15.00	18.00	21.00	24.0
*30	—	4.00	4.76	5.70	6.50	7.80	6.80	8.96	11.13	13.30	13.30	16.60	19.80	23.00	26.0
*36	—	4.60	5.30	6.20	7.00	8.64	7.30	9.63	11.96	14.30	14.30	17.80	21.30	24.80	28.0
Flat Surface	.73	.86	1.0*	1.18	1.37	1.83	1.39	1.54	1.90	2.19	2.23	2.55	2.84	3.12	3.39

*Factors for piping based upon material price for 2 inches on 2 inch pipe, factors for flat surfaces based upon material price for 2 inches on flat surface.

46

Table 4-5. Unit Material Selling Price Ratios

(Specification: Cellular Glass)

Pipe Size (inches)	Single Layer				Double Layer					Triple Layer			
	1	1-1/2	2	2-1/2	3	4	3	4	5	6	7	8	9	10
1/2	0.30	0.52	0.76	.96	1.15	2.00	1.21	2.00	2.87	3.70	—	—	—	—
3/4	0.32	0.54	0.80	1.00	1.20	2.05	1.31	2.16	2.97	3.80	—	—	—	—
1	0.35	0.57	0.85	1.05	1.25	2.15	1.40	2.30	3.10	3.95	—	—	—	—
1-1/2	0.40	0.64	0.93	1.14	1.35	2.25	1.50	2.40	3.25	4.15	—	—	—	—
2	0.46	0.70	1.0*	1.22	1.45	2.35	1.60	2.50	3.40	4.27	—	—	—	—
2-1/2	0.52	0.78	1.08	1.31	1.54	2.45	1.70	2.60	3.60	4.50	—	—	—	—
3	0.59	0.85	1.15	1.39	1.63	2.58	1.80	2.80	3.80	4.80	—	—	—	—
4	0.70	0.95	1.32	1.60	1.90	2.92	1.90	2.98	4.07	5.15	5.20	6.20	7.20	8.20	9.20
5	0.77	1.05	1.50	1.85	2.20	3.20	2.10	3.23	4.36	5.50	5.50	6.60	7.70	8.80	9.90
6	0.91	1.18	1.58	2.06	2.46	3.40	2.30	3.50	4.70	5.90	6.00	7.30	8.50	9.80	11.00
8	—	1.42	1.94	2.39	2.84	3.80	2.50	3.90	5.30	6.75	6.90	8.30	9.60	11.00	12.30
10	—	1.70	2.25	2.76	3.20	4.20	2.90	4.40	6.00	7.50	7.75	9.20	10.60	12.00	13.40
12	—	2.00	2.60	3.10	3.60	4.60	3.30	5.00	6.70	8.40	8.60	10.20	11.80	13.4	15.00
14	—	2.32	2.85	3.48	4.04	5.20	3.80	5.60	7.50	9.30	9.30	11.10	12.90	14.70	16.50
16	—	2.50	3.15	3.75	4.34	5.60	4.30	6.30	8.20	10.20	10.00	12.00	14.00	16.00	18.00
18	—	2.65	3.40	4.20	4.80	6.20	4.90	6.90	9.00	11.00	10.80	13.00	15.20	17.40	19.60
20	—	2.90	3.65	4.45	5.30	6.90	5.40	7.50	9.70	11.80	11.70	14.10	16.50	18.90	21.30
24	—	3.36	4.20	5.15	6.10	8.00	6.30	8.60	10.90	13.20	14.00	16.80	19.60	22.40	25.00
*30	—	3.90	4.90	6.00	7.10	9.30	7.70	10.30	12.90	15.50	17.00	20.00	23.00	26.00	29.00
*36	—	4.60	5.70	7.00	8.20	10.50	9.20	12.00	14.70	17.50	20.00	24.00	28.00	32.00	36.00
Flat Surface	.68	.84	1.0*	1.16	1.32	1.64	1.36	1.68	2.04	2.36	2.4	2.7	3.0	3.3	3.6

*Factors for piping based upon material price for 2 inches on 2 inch pipe
 factors for flat surfaces based upon material price for 2 inches on flat surface.

Table 4-6. **Material Type Correction Factors, F_t**

Material	Correction Factor, F_t Piping	Flat Surfaces
Calcium Silicate	1.00	1.00
Preformed Fiber Glass	.85	.90
Preformed Mineral Wool	.90	.95
Cellular Glass	1.15	1.10

Notes. The above materials are installed with 0.016 aluminum jacket. Increase in cellular glass factors 15 to 20 percent for cold service joint sealing.

Table 4-7. **Regional Labor Productivity Factors, F_r**

City or State		City or State		City or State	
Alabama	0.8	Kansas	1.0	Ohio	1.0
Alaska	1.4	Kentucky	1.0	Oklahoma	0.9
Arizona	1.1	Louisiana	0.9	Oregon	1.2
Arkansas	0.9	Maine	1.1	Pennsylvania	1.0
California	1.0	Maryland	1.2	Rhode Island	1.1
Colorado	0.9	Massachusetts	1.1	S. Carolina	1.0
Connecticut	1.1	Michigan	1.0	S. Dakota	1.0
Delaware	0.9	Minnesota	1.0	Tennessee	1.0
District of Columbia	1.2	Mississippi	0.9	Texas	0.8
		Missouri	1.0	Utah	1.0
Florida	0.9	St. Louis	1.2	Vermont	1.1
Miami	1.0	Montana	1.2	Virginia	0.9
Georgia	0.8	Nebraska	1.0	Washington	1.2
Atlanta	1.0	Nevada	1.2	W. Virginia	1.0
Hawaii	1.2	New Hampshire	1.1	Wisconsin	1.0
Idaho	1.0	New Jersey	1.0	Wyoming	1.0
Illinois	1.1	New Mexico	0.8	Ontario	1.2
Chicago	1.2	New York	1.1	Quebec	1.2
Indiana	1.1	New York City	1.2	Maritime Provinces	1.4
Iowa	1.0	N. Carolina	1.0		
		N. Dakota	1.0		

Table 4-8. Base Worker Productivity, WP

Piping - man-hours/100 linear feet of pipe
Flat surfaces - man-hours/100 sq.ft.

Pipe Size (inches)	Single Layer						Double Layer						Triple Layer			
	1	1-1/2	2	2-1/2	3	4	3	4	5	6	6	7	8	9	10
1/2	11.0	12.0	13.0	14.0	15.0	17.0	20.5	25.0	29.5	34.0	–	–	–	–	–
3/4	11.0	12.0	13.0	14.0	15.0	17.0	20.5	25.0	29.5	34.0	–	–	–	–	–
1	11.2	12.3	13.4	14.5	15.6	17.8	21.0	25.5	30.0	34.5	–	–	–	–	–
1-1/2	11.6	12.7	13.9	15.0	16.2	18.5	21.5	26.0	30.5	35.0	–	–	–	–	–
2	12.2	13.4	14.6	15.8	17.0	19.4	22.8	27.5	32.3	37.0	–	–	–	–	–
2-1/2	12.9	14.2	15.5	16.8	18.0	20.6	24.0	28.0	33.0	39.0	–	–	–	–	–
3	13.7	15.0	16.5	17.8	19.2	22.0	25.0	30.0	35.5	41.0	–	–	–	–	–
4	14.5	16.1	17.7	19.2	20.8	24.0	26.5	32.0	37.5	43.0	50	57	63	70	77
5	15.5	17.3	19.0	20.8	22.6	26.2	29.0	34.5	40.0	45.5	54	61	68	75	82
6	17.0	18.9	20.8	22.7	24.7	28.5	32.0	37.5	43.5	48.4	58	65	73	80	87
8	–	21.0	23.0	25.0	27.0	31.0	35.6	41.4	47.2	53.0	66	73	81	89	95
10	–	23.0	25.2	27.4	29.6	34.0	38.8	45.4	52.0	58.7	74	81	89	96	103
12	–	25.0	27.4	29.8	32.2	37.0	43.0	50.0	57.0	64.0	82	89	97	104	111
14	–	27.6	30.1	32.6	35.1	40.5	49.5	57.0	64.5	72.0	90	98	106	114	122
16	–	30.9	33.7	36.5	39.4	45.0	56.0	64.0	72.0	80.0	98	106	115	123	131
18	–	34.2	37.2	40.1	43.0	49.0	59.5	68.0	76.5	85.0	108	117	126	135	144
20	–	38.5	41.6	44.7	47.8	54.0	63.0	73.0	83.0	93.0	120	130	140	150	160
24	–	43.0	46.8	50.6	54.4	62.0	72.0	85.0	98.0	110.0	140	153	167	180	193
*30	–	60.0	65.0	70.0	75.0	85.0	95.0	110.0	125.0	140.0	165	185	205	225	245
*36	–	70.0	78.0	86.0	94.0	110.0	118.0	135.0	153.0	170.0	190	212	234	256	278
Flat Surface	12.2	13.0	13.8	14.6	15.4	16.2	20.5	22.0	23.5	25.0	29.0	30.3	31.6	33	34.3

Insulation Thickness, Inches

*Insulation applied in quarter segments.

Note. All productivity data is based upon installing calcium silicate (on piping only) on a 100 foot straight run of pipe with a fitting or obstruction every 5 feet. Piping is 8 feet from working level and is accessible with a 6 foot step ladder. Work is located outdoors during summer conditions. Worker efficiency is based on that of a 40 hour man/week.

49

CHAPTER 5
ECONOMIC THICKNESS DETERMINATION

Introduction

The final procedure in calculating the economic thickness of an insulation material is presented in this section. The solution is valid for either hot or cold systems. The cost of energy determined in Chapter 3 is required as is the marginal cost of insulation determined in Chapter 4. The equations used in this section are developed in Appendix A.

Contents

1. Glossary
2. Economic Thickness Determination Worksheet
3. Nomographs

Glossary

B_3 annual amortization factor to amortize the cost of insulation

Btu   British thermal unit

D_p annual cost of heat lost through one linear foot of pipe insulation

D_{pr} annual cost of heat gain through one linear foot of pipe insulation

D_s annual cost of heat lost through one ft^2 flat surface insulation

D_{sr} annual cost of heat gained through one ft^2 flat surface insulation

$°F$ degrees Fahrenheit

ft^2 square feet

hr   60 minutes

i_3 cost of insulation rate or required rate of return on the last increment of insulation

in   inches

k	insulation thermal conductivity, $\dfrac{\text{Btu-in}}{\text{hr-ft}^2\text{-}°F}$
M	project average annual cost of heat, $/10^6$ Btu
M_r	project average annual cost of refrigeration, $/10^6$ Btu
m_c	incremental cost of installed insulation, $/in
m_{c1}	incremental cost of installed insulation, single layer, $/in
m_{c2}	incremental cost of installed insulation, double layer, $/in
m_{c3}	incremental cost of installed insulation, triple layer, $/in
n_1	term of insulation project, yrs.
R_s	surface resistivity of the insulation system, $\dfrac{\text{hr-}°F\text{-ft}^2}{\text{Btu}}$
ΔT	$t_p - t_a$, difference between pipe or flat surface temperature and ambient temperature, $°F$
t_a	average ambient temperature, $°F$ (design temperature for application)
t_p	process temperature, $°F$ (temperature of surface to be insulated)
w_1	economic thickness of insulation, in, single layer application
w_2	economic thickness of insulation, in, double layer application
w_3	economic thickness of insulation, in, triple layer application
Y	annual operating time for insulation, hr
Z	factor used in economic thickness determination, $Z = \dfrac{D}{1.1\, m_c B_3}$
$	dollars

Economic Thickness Determination Worksheet
Hot and Cold* Systems

1. Calculate Mean Insulation Temperature.

    $$\frac{t_p + t_a}{2}$$ $t_m=$ _____ °F

2. Enter Insulation Thermal conductivity using Figs. 5-1 or 5-2. $k =$ _____ $\frac{Btu- n}{hr-ft^2-°F}$

3. Calculate Temperature Difference (t_p-t_a). $\Delta T=$ _____ °F

4. Enter Annual Hours of Operation. $Y=$ _____ hrs

5. Find D_s for flat or D_p for pipe using Fig. 5-3. $D_s=$ _____ $D_p=$ _____

6. Determine B_3 using Fig. 5-4 and 5-5 using i_3 money cost for insulation and n_1 insulation project life. $i_3=$ 0. _____ $n_1=$ _____ yrs $B_3=$ _____

7. Use m_{cl} from Chapter 4. $m_{cl}=$ _____

8. Find Z_s using Fig. 5-6 for flat or Z_p using Fig. 5-7 for pipe. $Z_s=$ _____ $Z_p=$ _____

9. Calculate kR_s k x 0.7 (R_s value) of 0.7 is typical) $kR_s=$ _____

10. Use proper Fig. 5-8 to determine economic thickness. $w_1=$ _____

*If a cold system, see Chapter 6 also.

Flat	_____
1/2"x	_____
3/4"x	_____
1" x	_____
1-1/2"x	_____
2" x	_____
2-1/2"x	_____
3" x	_____
4"x	_____
5"x	_____
6"x	_____
8"x	_____
10"x	_____
12"x	_____
14"x	_____
16"x	_____
18"x	_____
20"x	_____
24"x	_____
30"x	_____
36"x	_____

11. If the economic thickness found in Step 10 is within the single layer range (corresponding to the single layer slope, m_{c1}, used in Step 7), the thickness is correct. If the thickness is beyond the single layer range, repeat the procedure from Step 7 on, using the double layer slope, m_{c2}.

$w_2 = $ _____

12. If the economic thickness, using the double layer slope, is in the triple layer range, repeat the procedure from Step 7 on, using the triple layer slope, m_{c3}.

$w_3 = $ _____

Note. It may happen that the economic thickness found with a single layer slope falls in the double layer realm and that the subsequent thickness found with a double layer slope falls in the single layer realm; or that the economic thickness found with a double layer slope falls in the triple layer realm and the subsequent thickness found with a triple layer slope falls in the double layer realm. Should either of these conditions occur, the proper economic choice is the thickest single or thickest double layer, respectively.

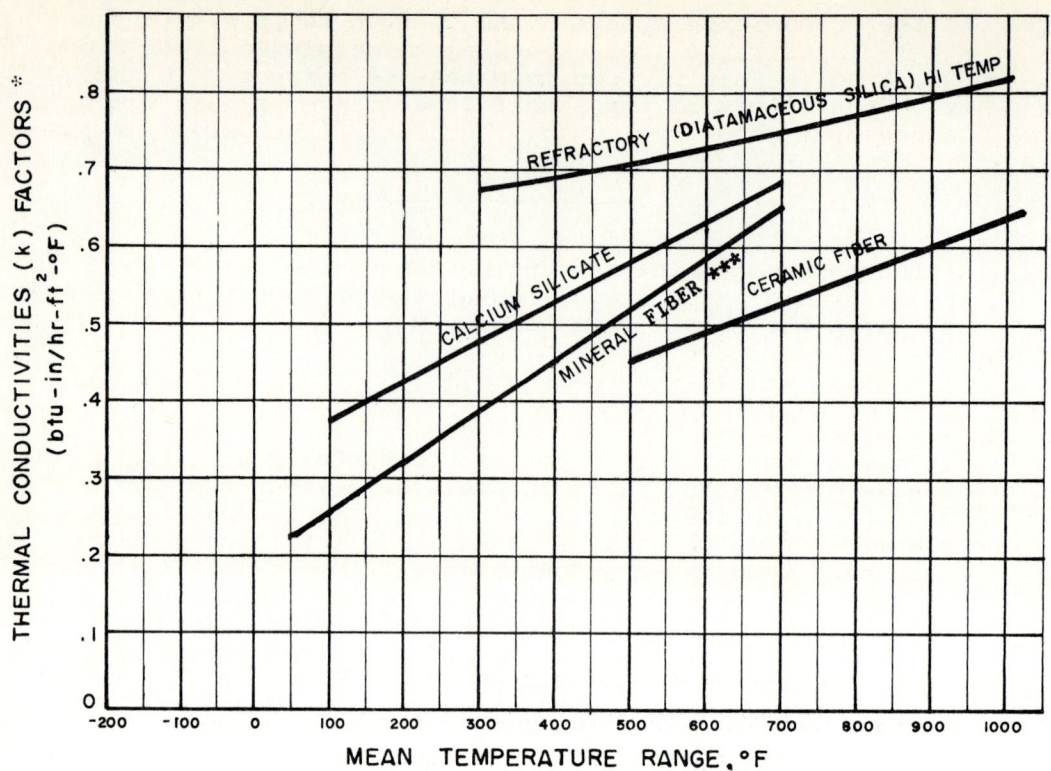

Figure 5-1. High temperature insulation materials

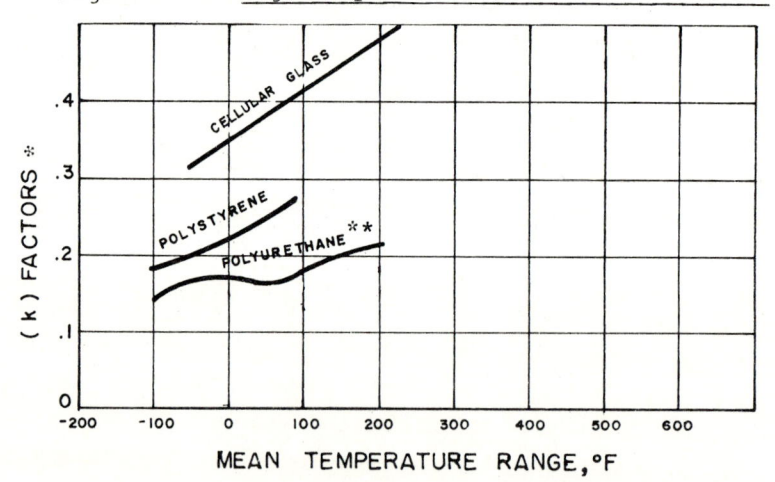

Figure 5-2. Low temperature insulation materials

*The determination of in-service values has yet to be determined for most materials. Should the user have accurate k values for his choice of material those values should be used.
**New material values. Increase by 25 percent for aged value.
***Mineral fiber means rock wool, slag wool, and glass fiber.

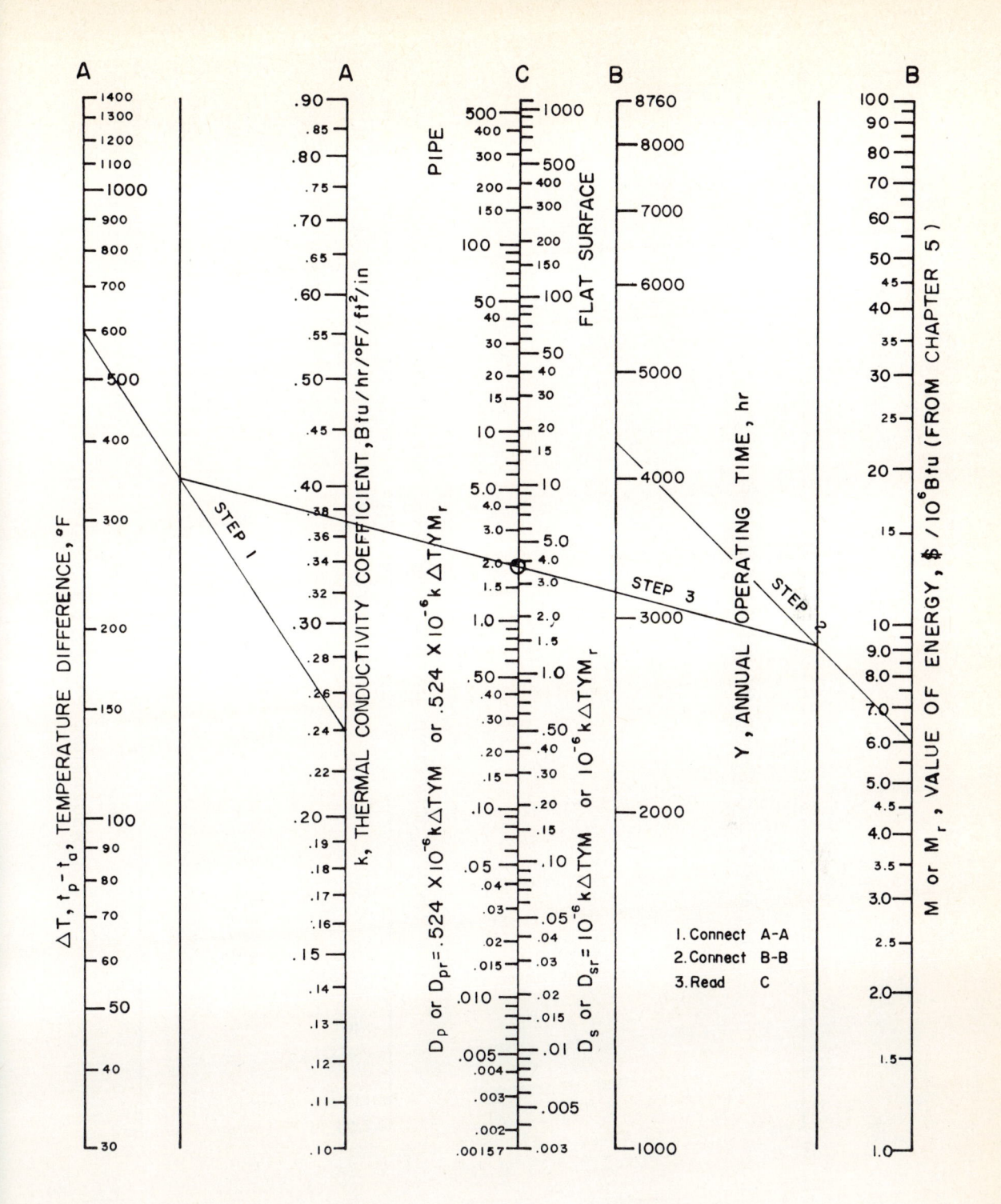

D_p, D_{pr}, D_s AND D_{sr} VALUES

Figure 5-3. Annual cost of heat lost or gained

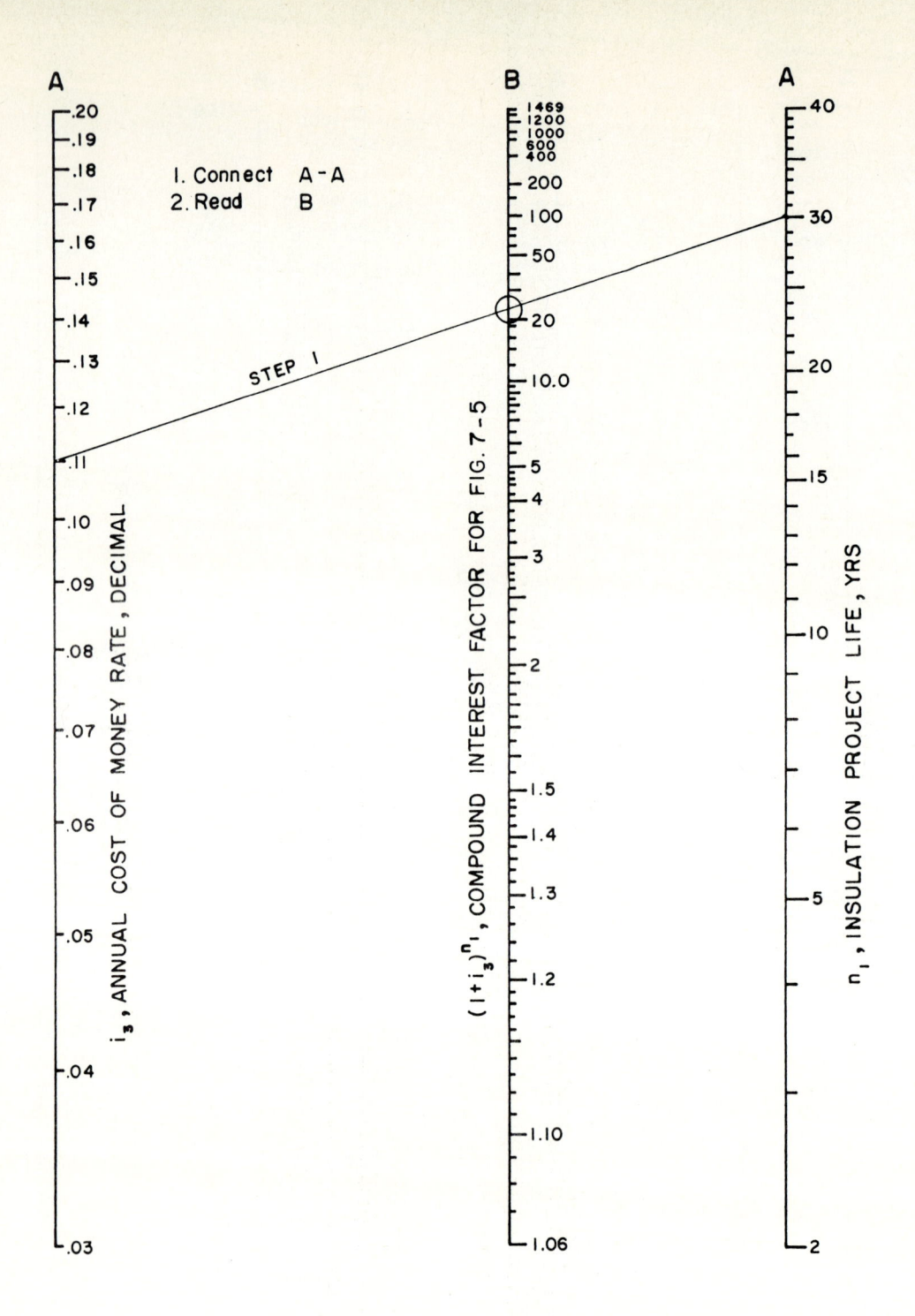

Figure 5-4. Compound interest factor for use with Figure 5-5

1. CONNECT A-A
2. READ B

STEP 1

B_3, MULTIPLIER

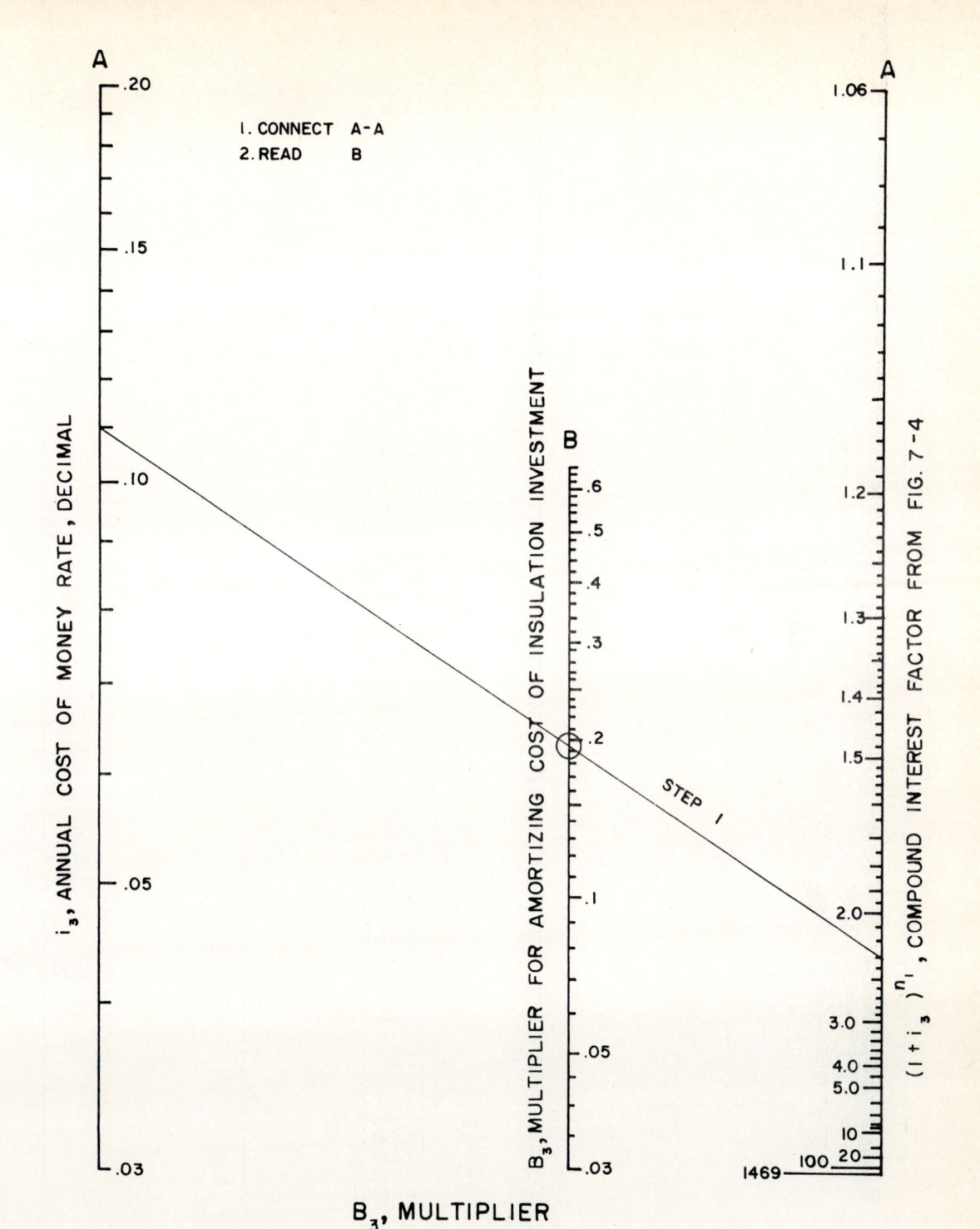

The B_3 multiplier $\frac{i(1+i)^n}{(1+i)^n - 1}$ for amortizing an initial cost over a period of time with equal increments.

Figure 5-5. The amortizing multiplier, B

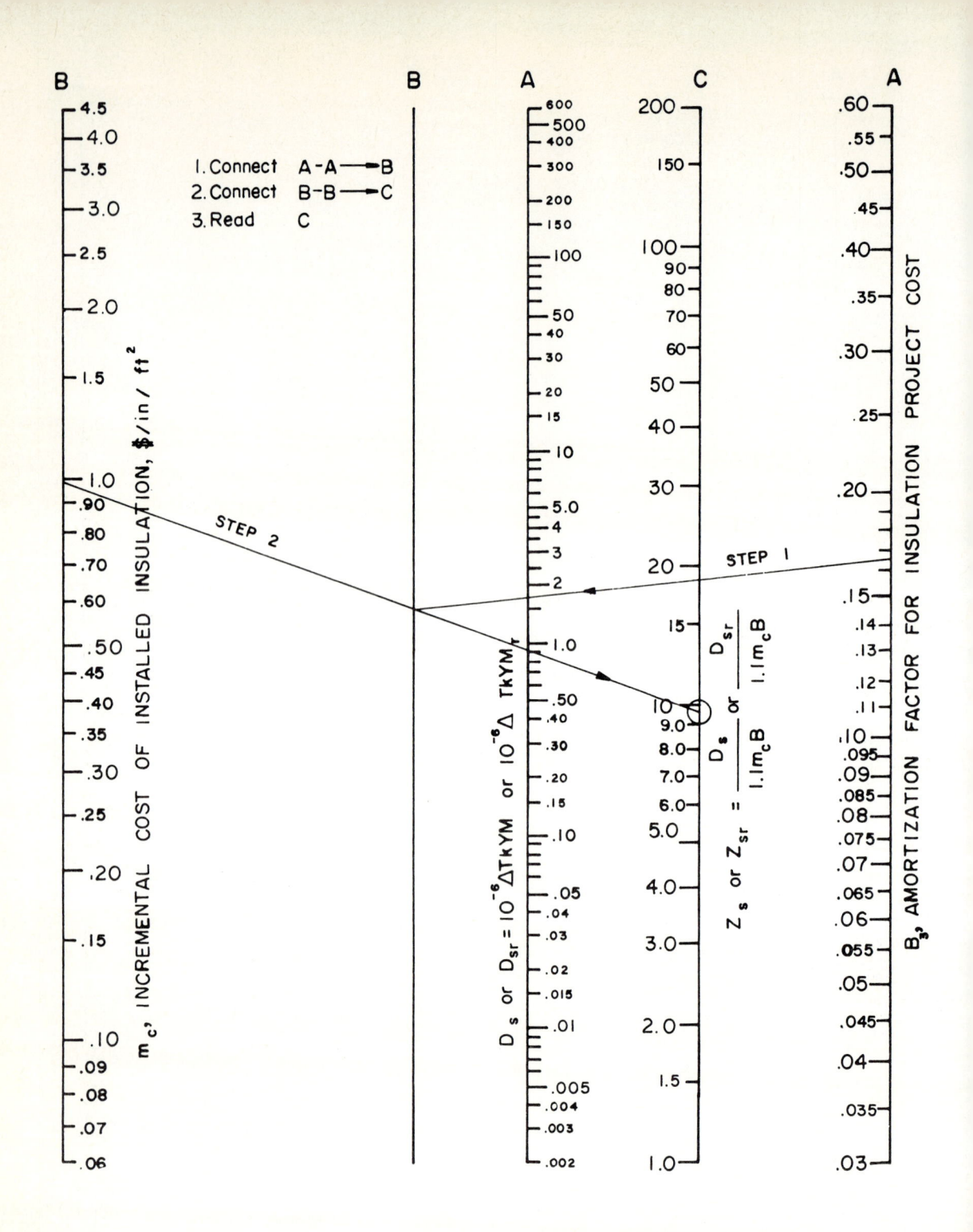

Note.— Z_s includes a 10 percent annual maintenance charge for insulation. See Appendix A.

Figure 5-6. Z_{s}, factor for flat surfaces

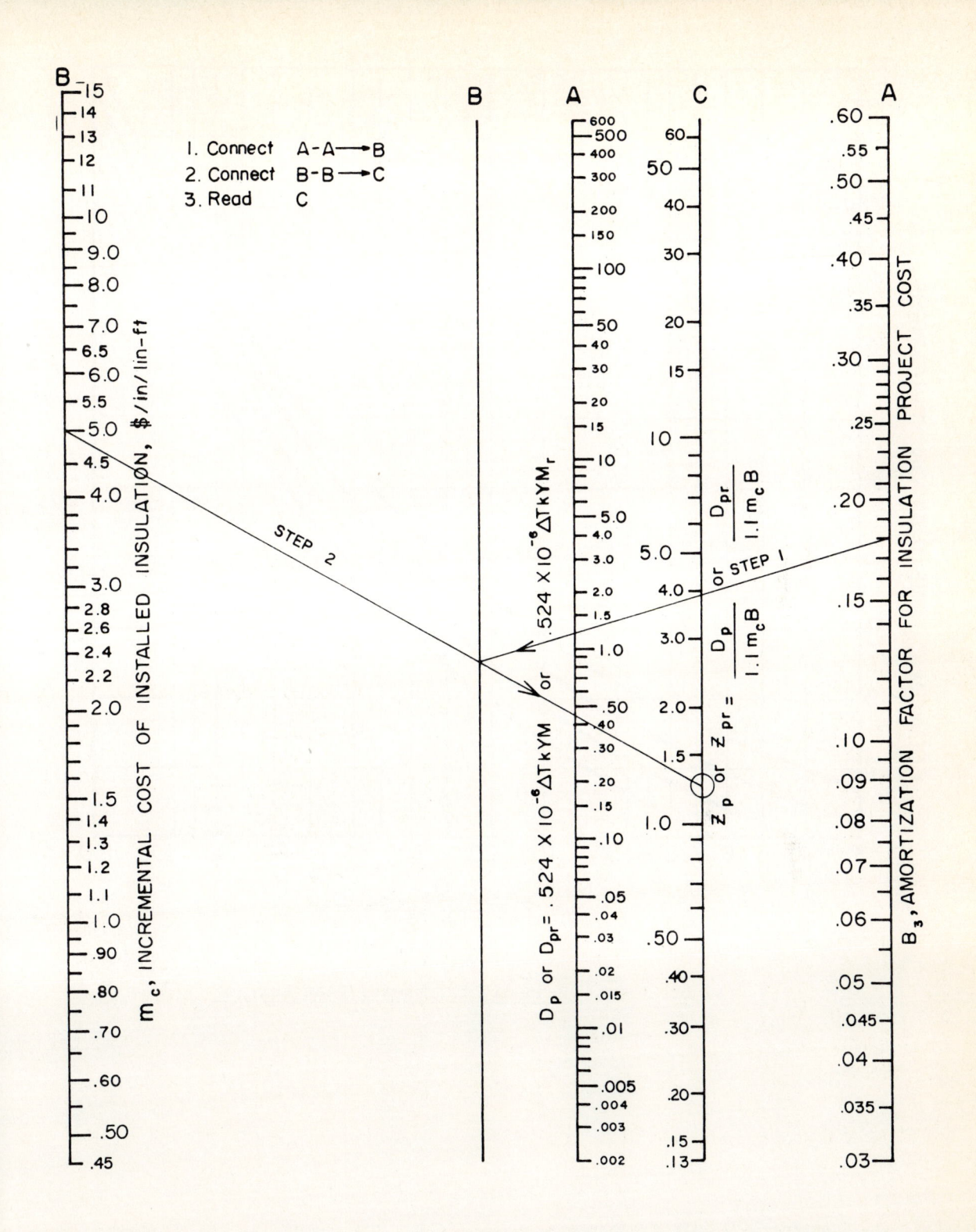

Note.— Z_p includes 10 percent annual maintenance charge for insulation. See Appendix A.

Figure 5-7. Z_p, factor for round (pipe) surfaces

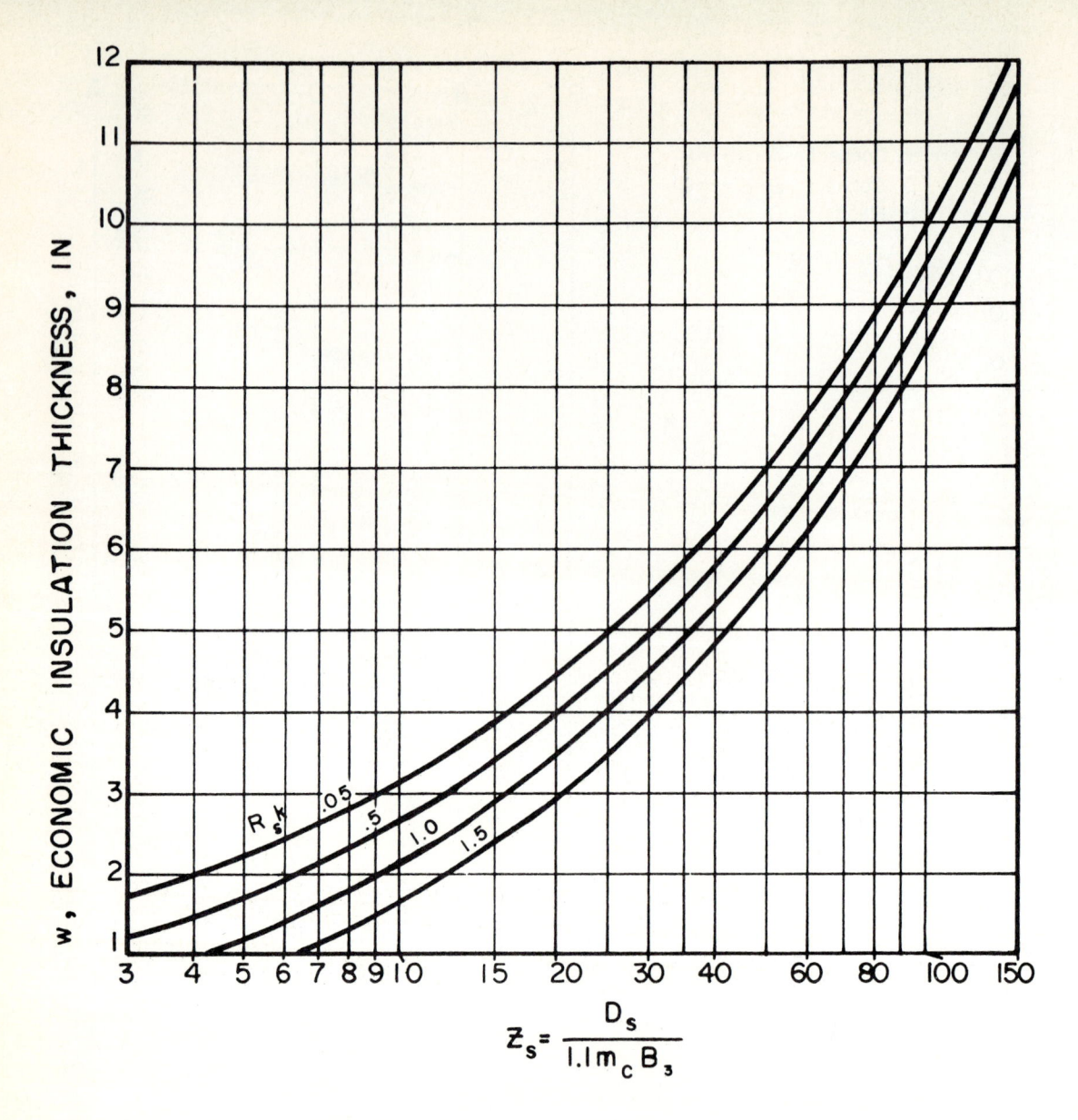

Figure 5-8a. Economic thickness, flat surface

$$Z_s = \frac{D_s}{1.1 m_c B_3}$$

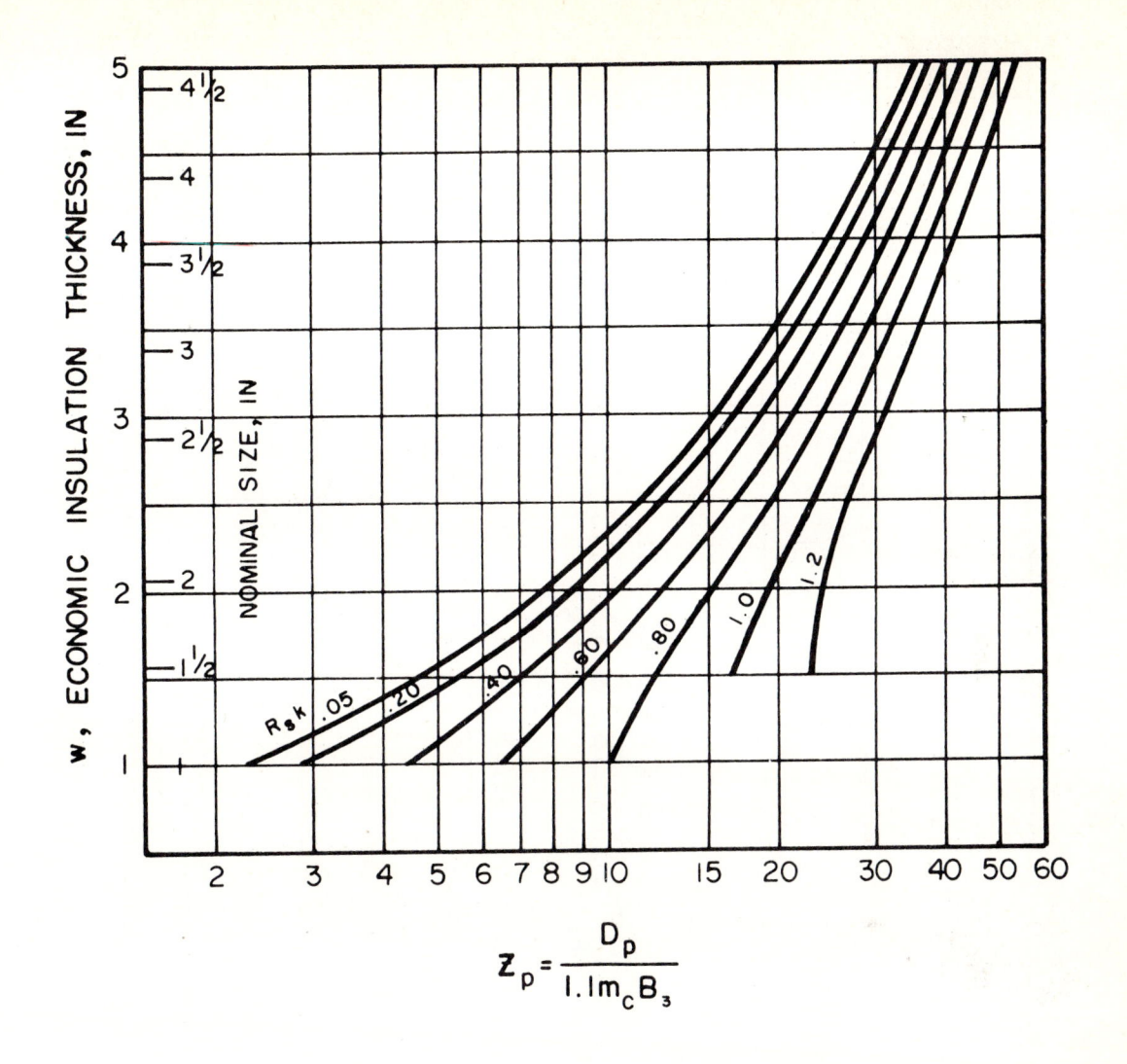

Figure 5-8b. w, Economic thickness, ½-inch pipe

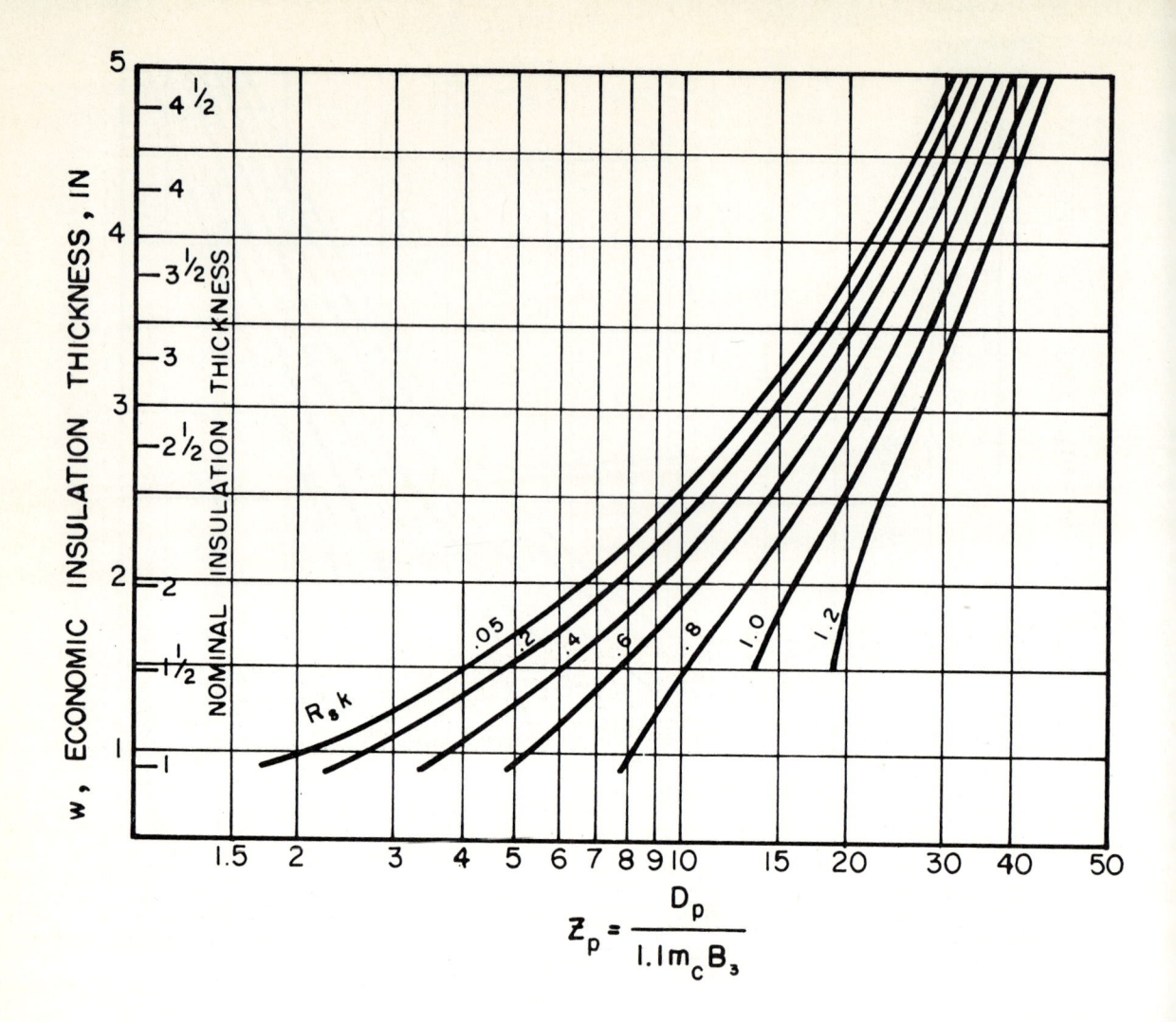

Figure 5-8c. w, Economic thickness, 3/4-inch pipe

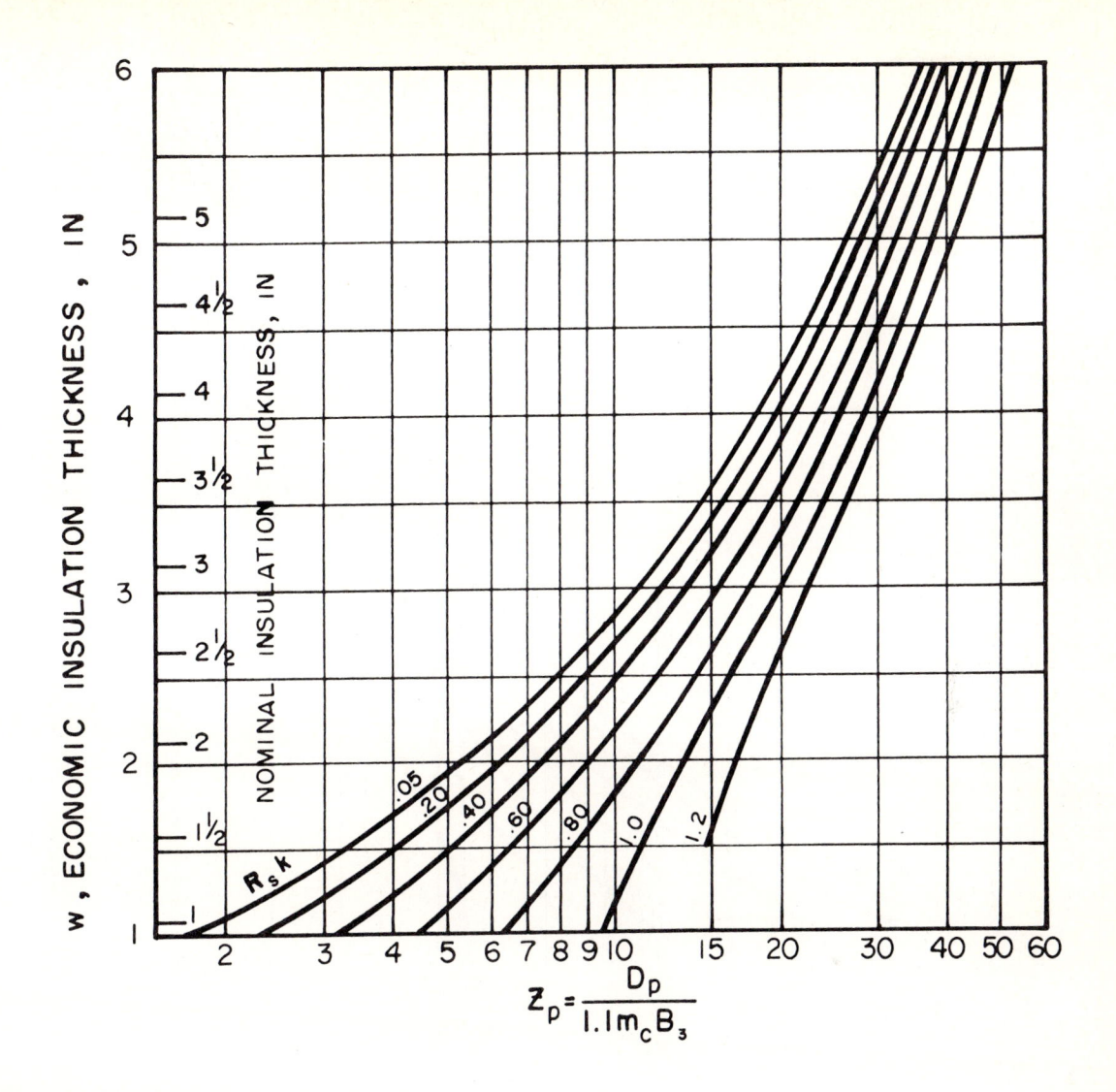

Figure 5-8d. w, Economic thickness, 1-inch pipe

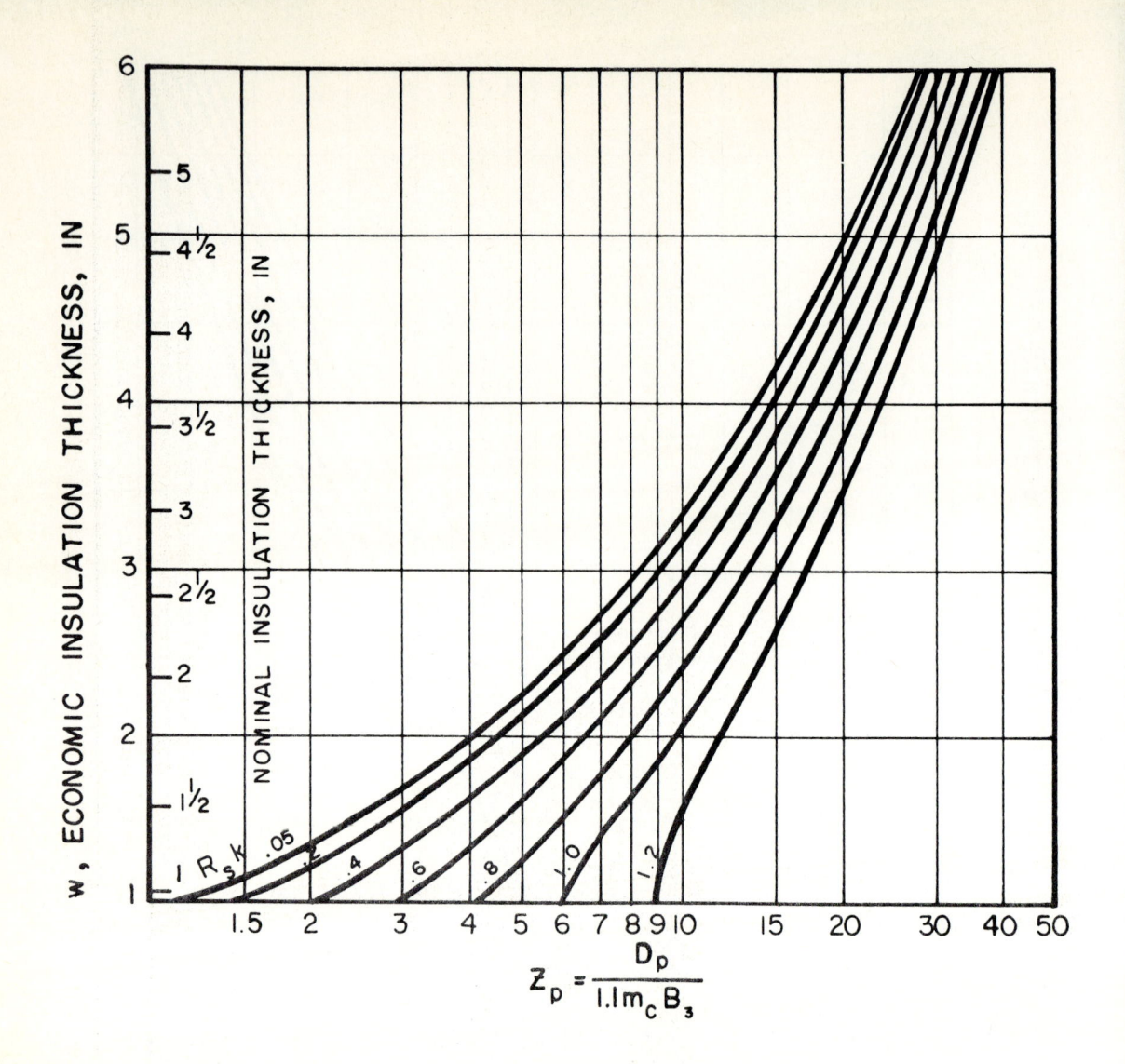

Figure 5-8e. w, Economic thickness, 1½-inch pipe

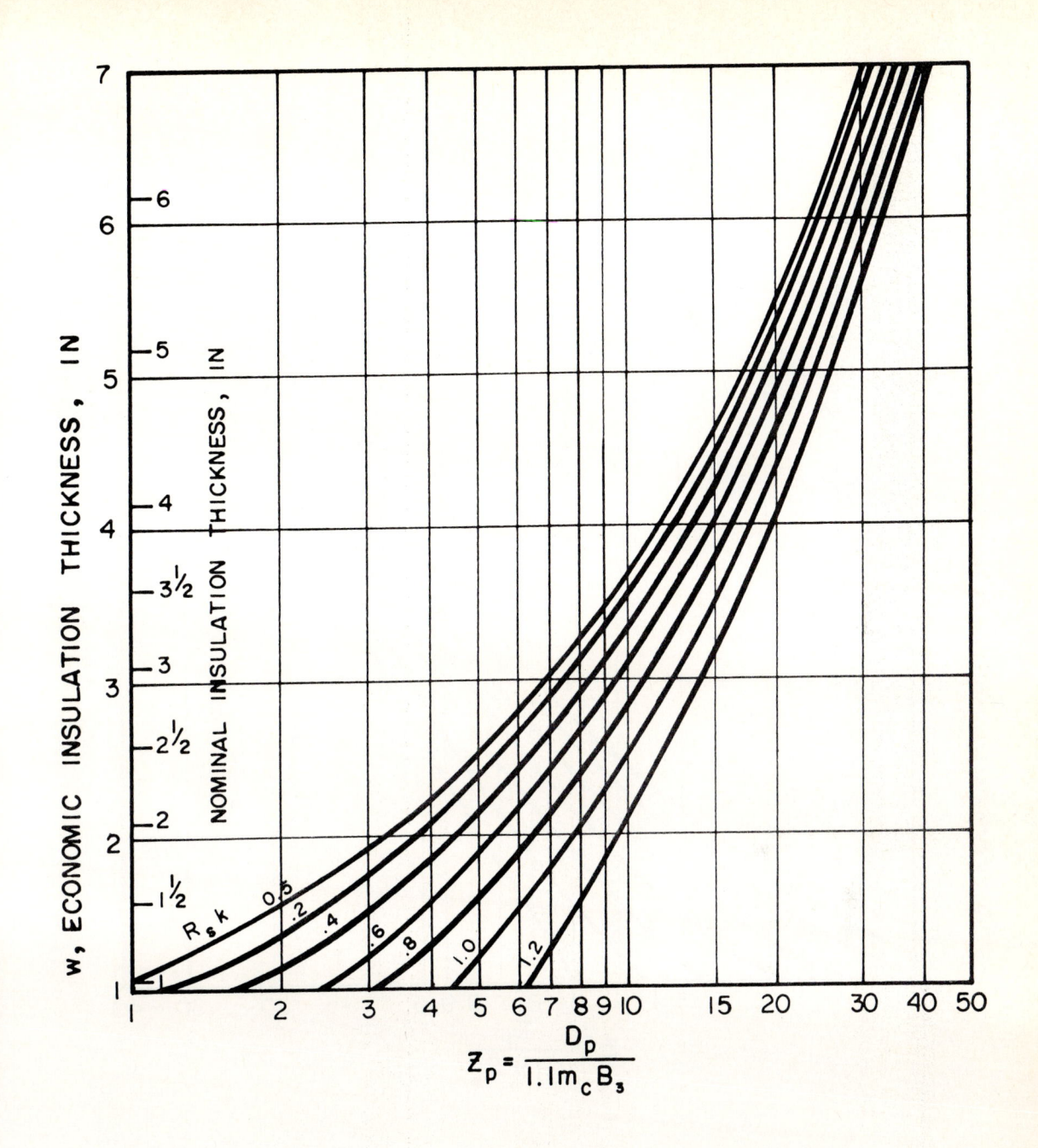

Figure 5-8f. w, Economic thickness, 2-inch pipe

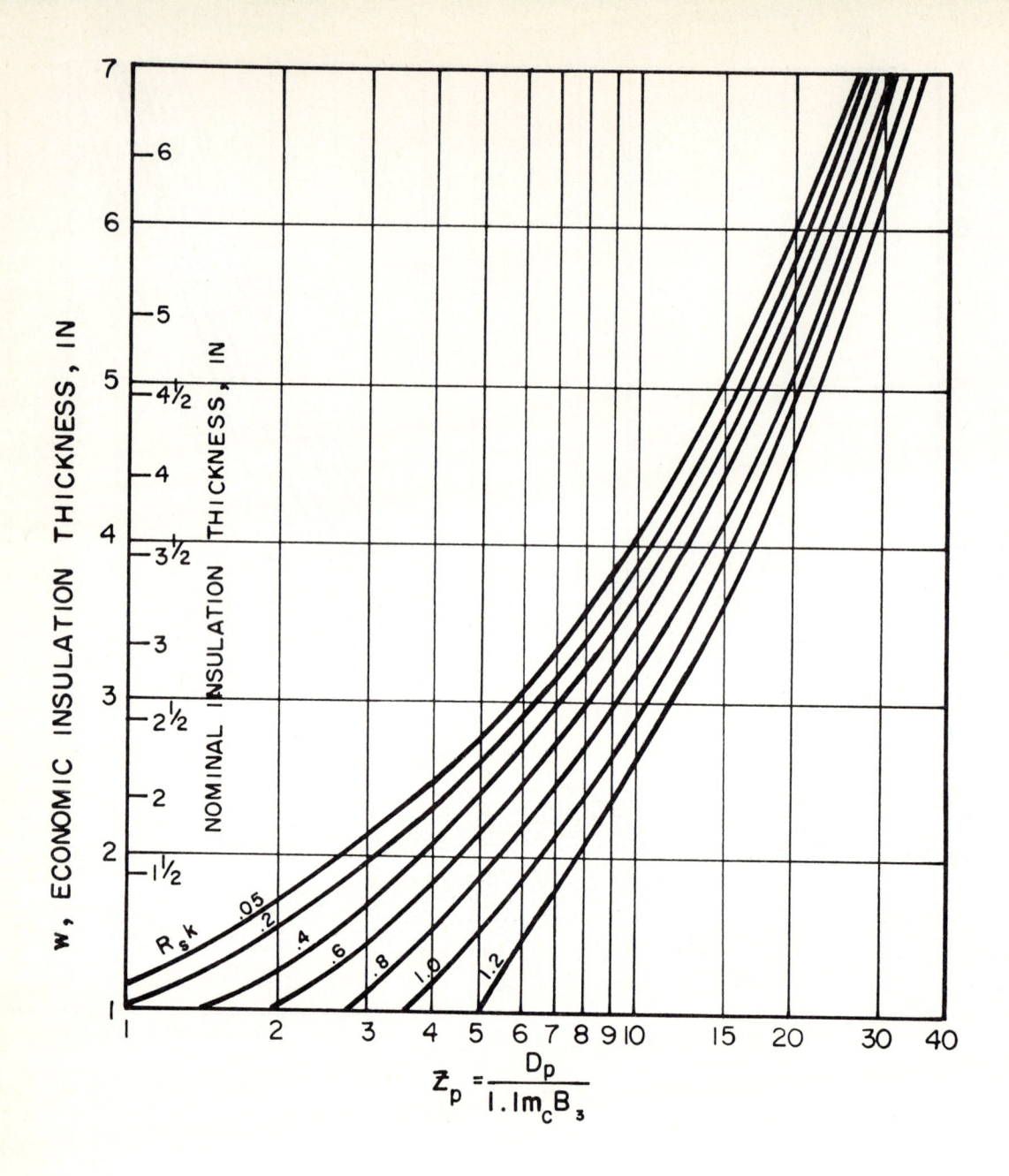

Figure 5-8g. w, Economic thickness, 2½-inch pipe

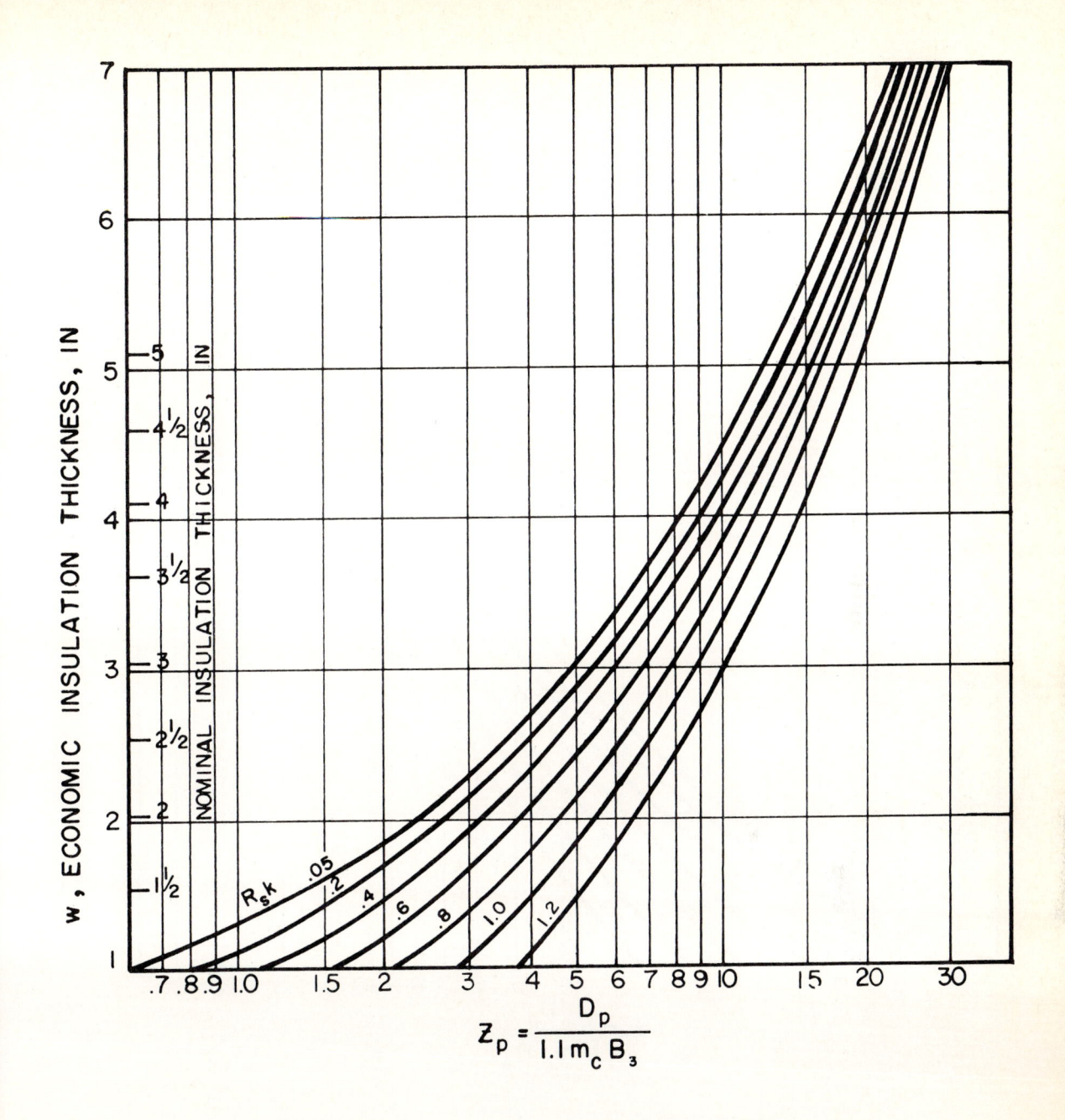

Figure 5-8h. w, Economic thickness, 3-inch pipe

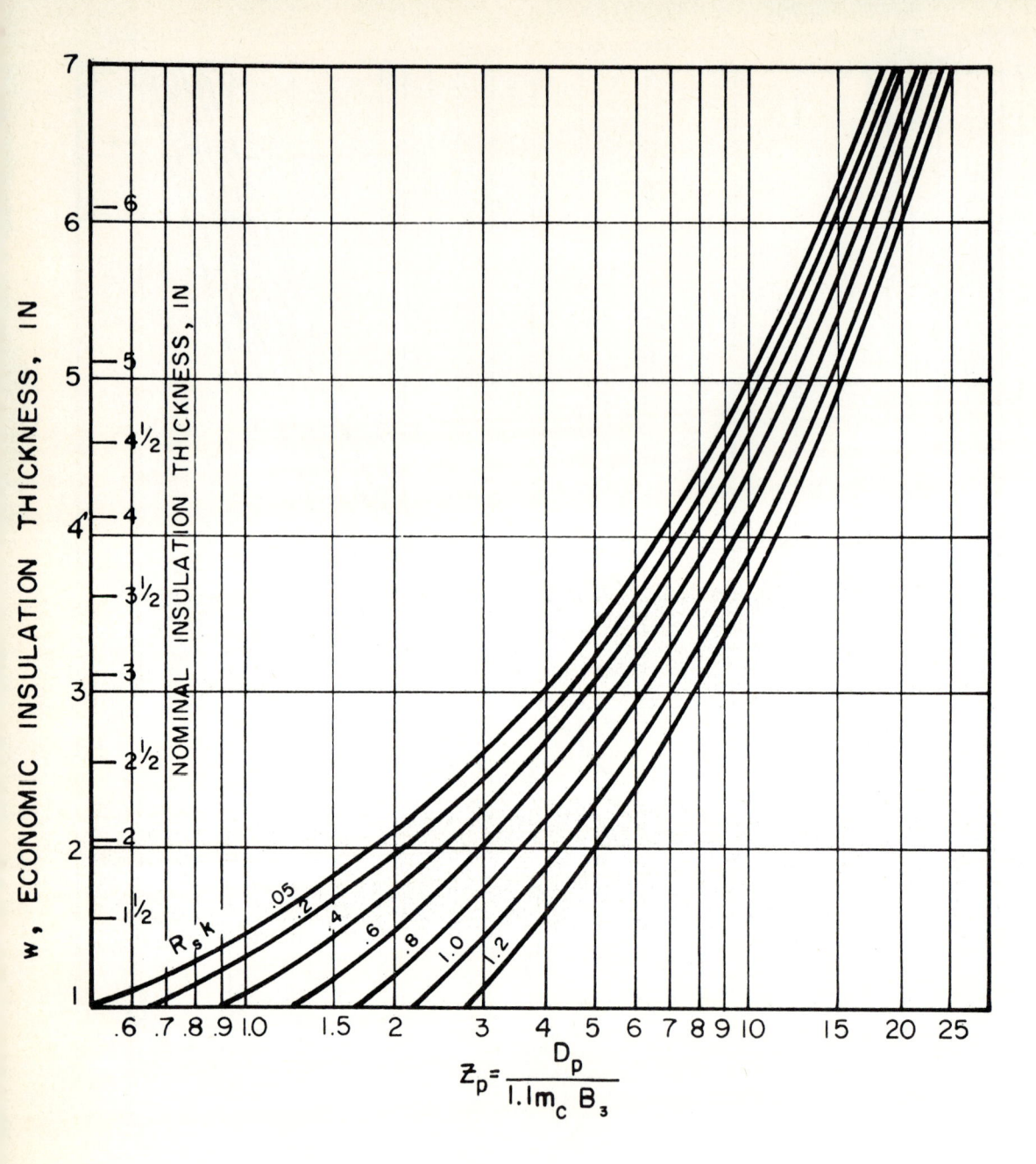

Figure 5-8i. w, Economic thickness, 4-inch pipe

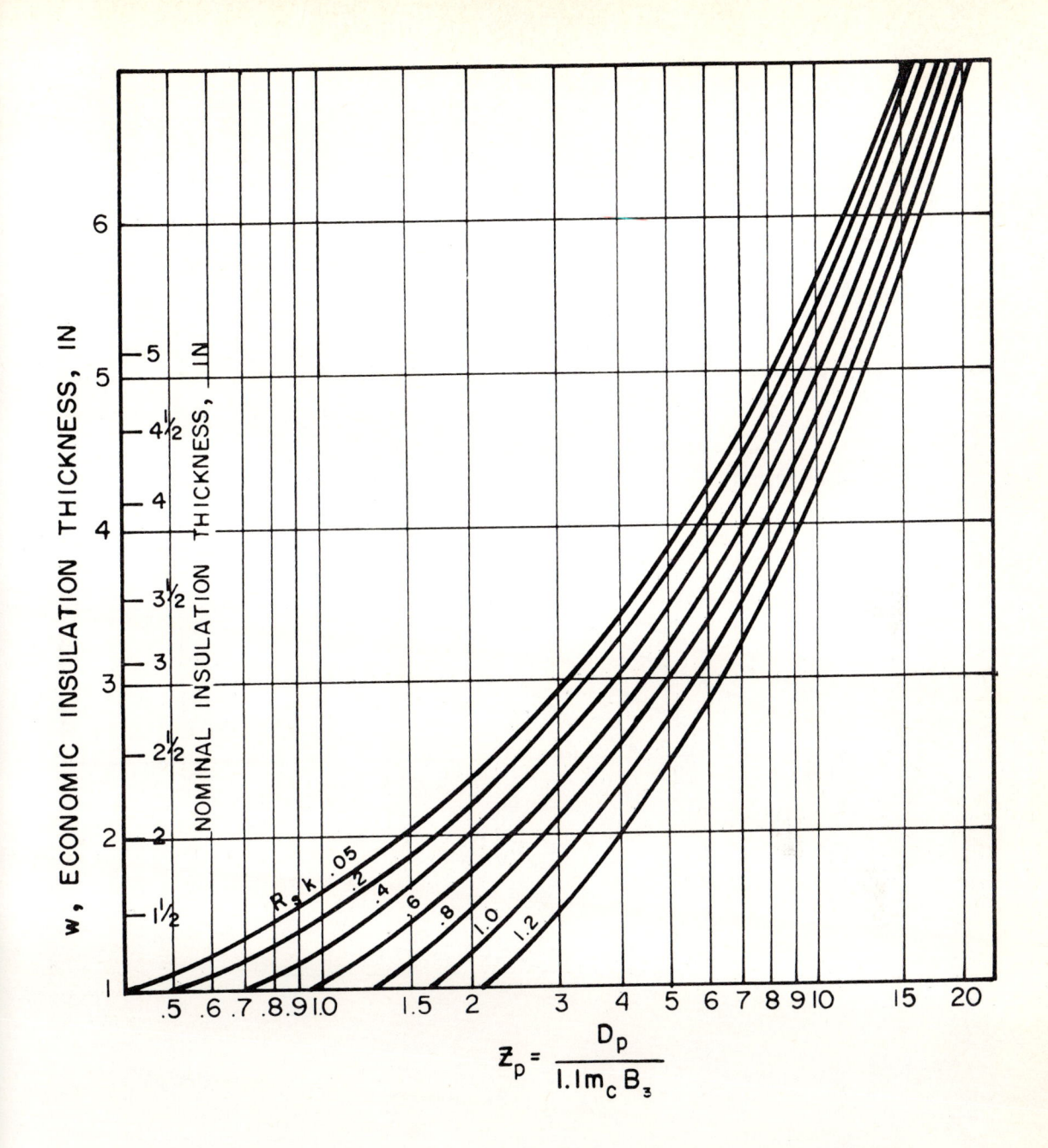

Figure 5-8j. w, Economic thickness, 5-inch pipe

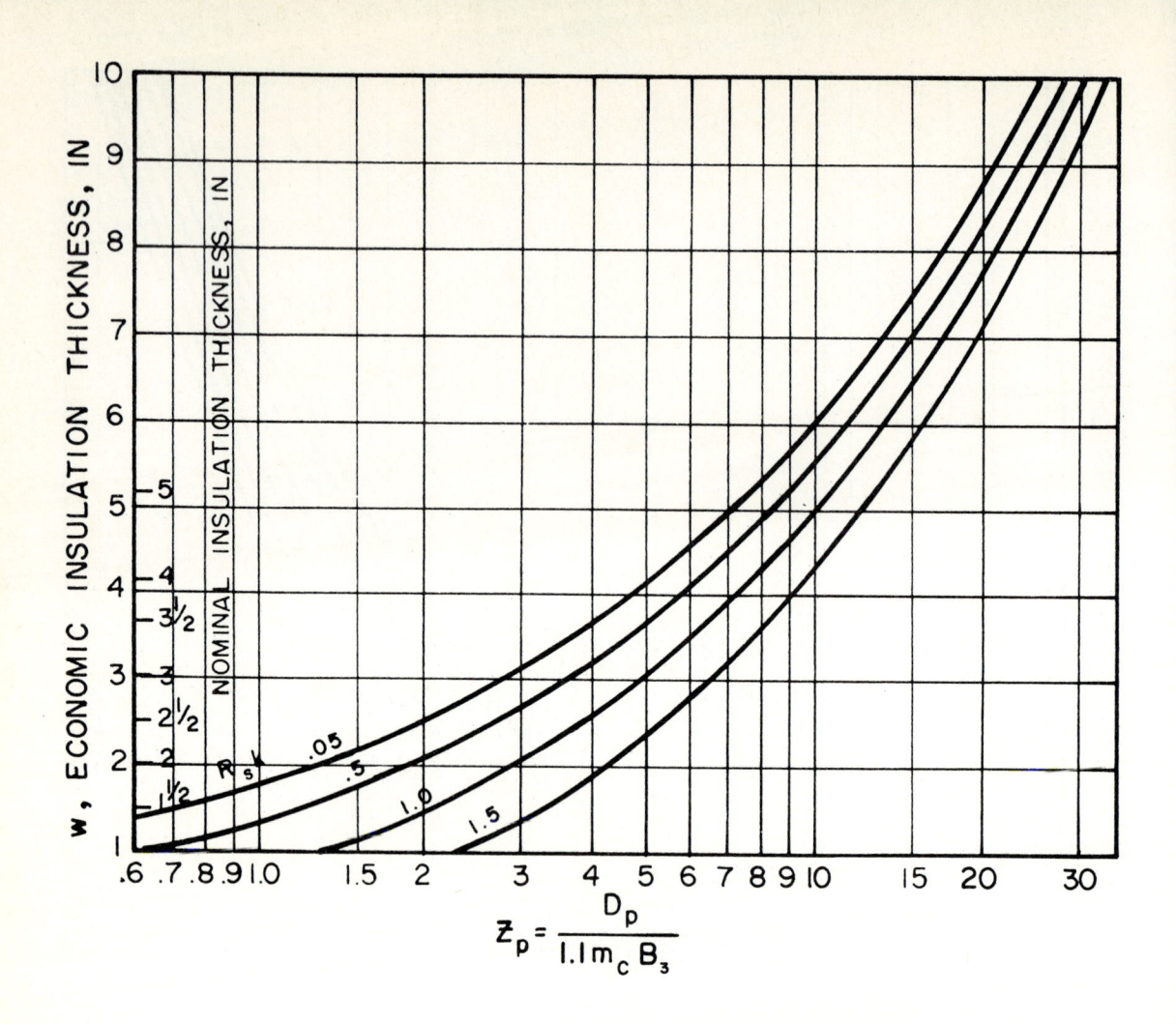

Figure 5-8k. w, Economic thickness, 6-inch pipe

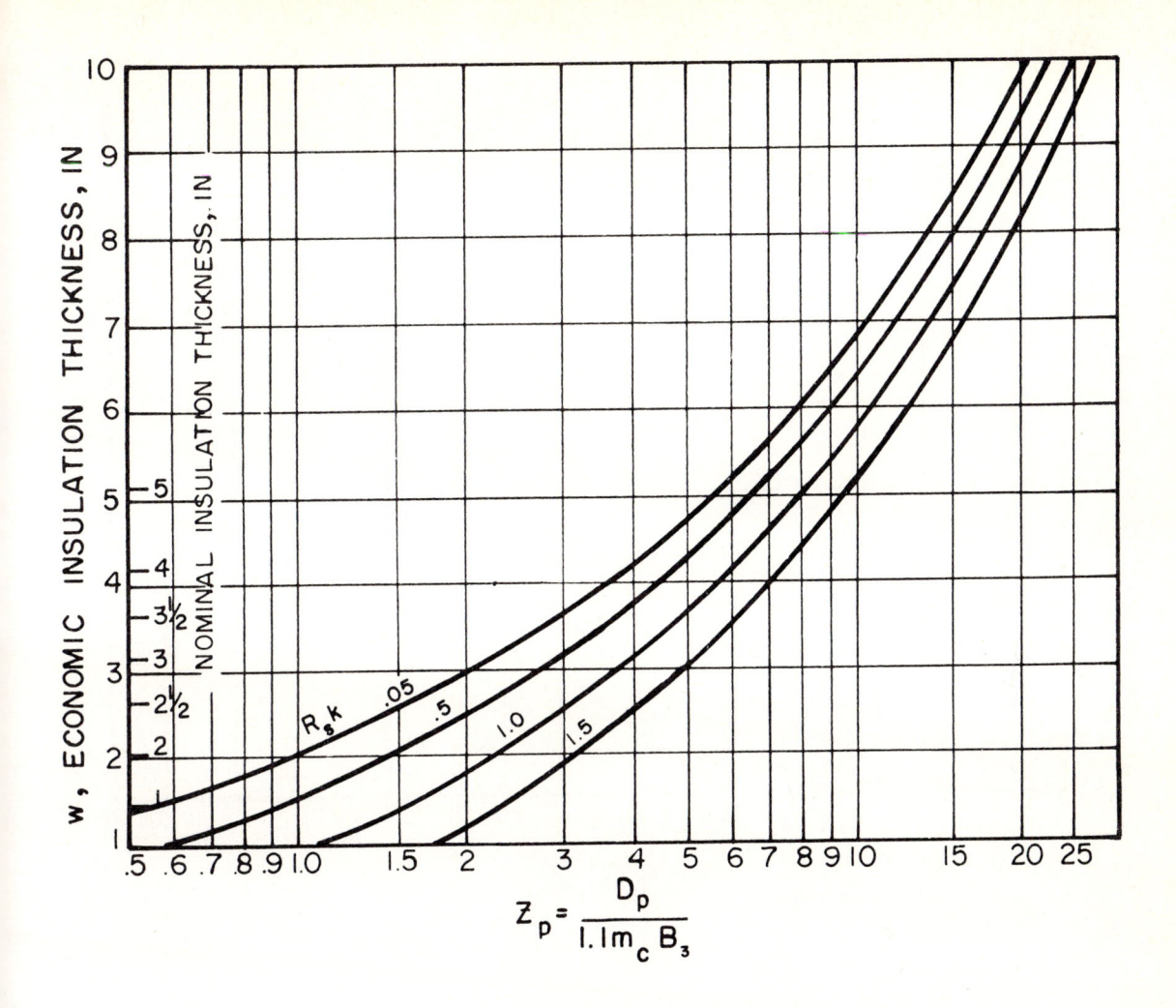

Figure 5-81. w, Economic thickness, 8-inch pipe

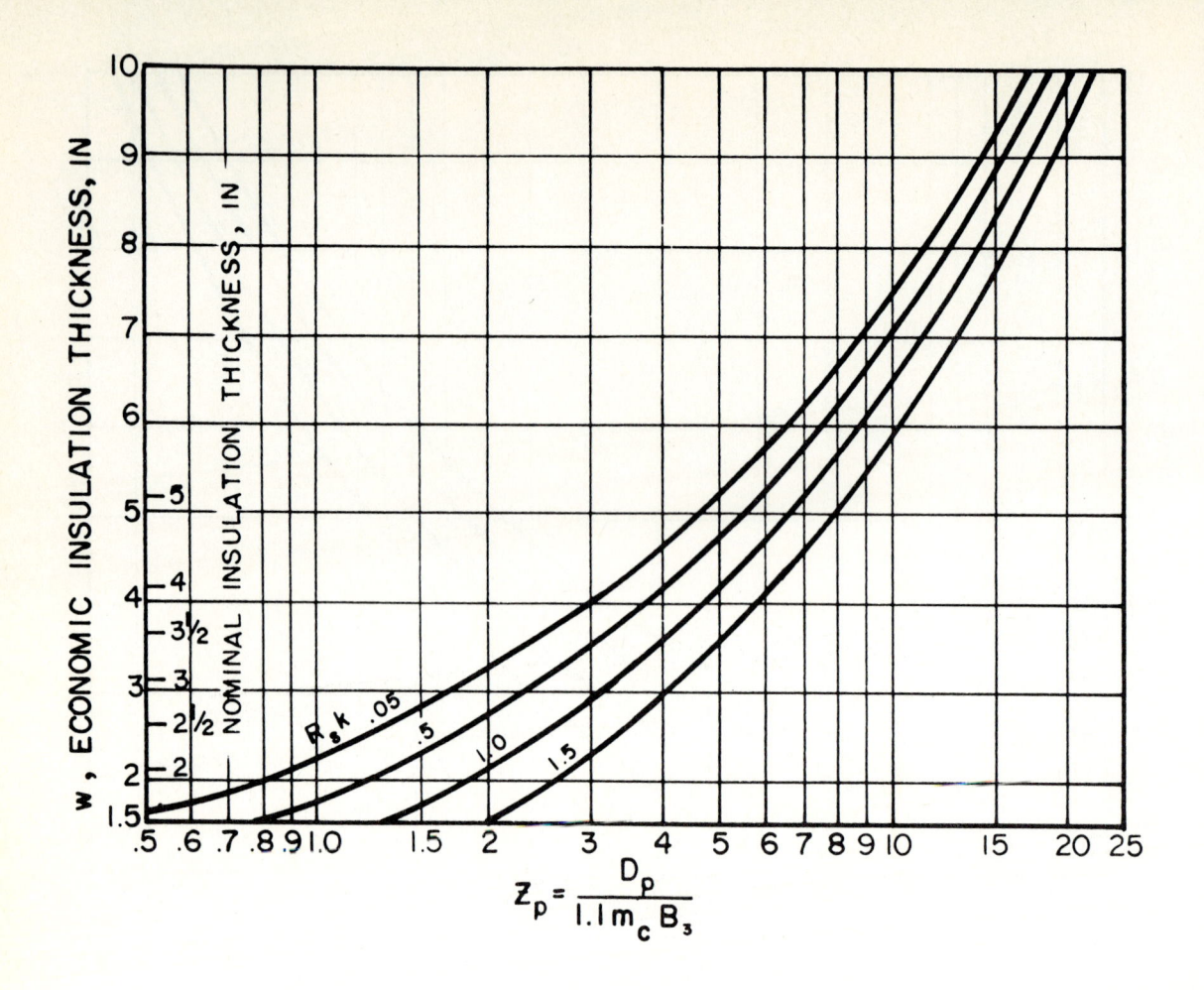

Figure 5-8m. w, Economic thickness, 10-inch pipe

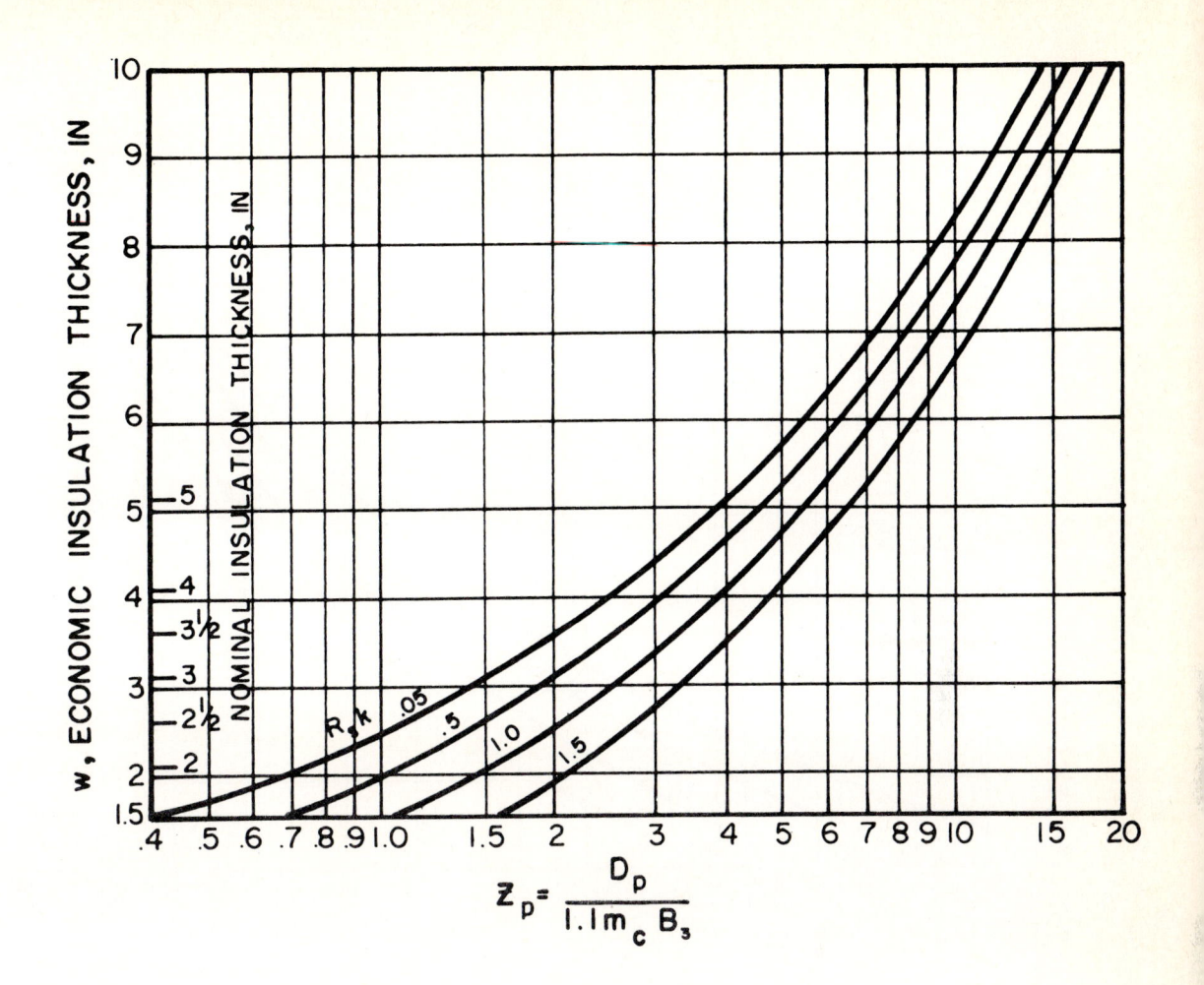

Figure 5-8n. w, Economic thickness, 12-inch pipe

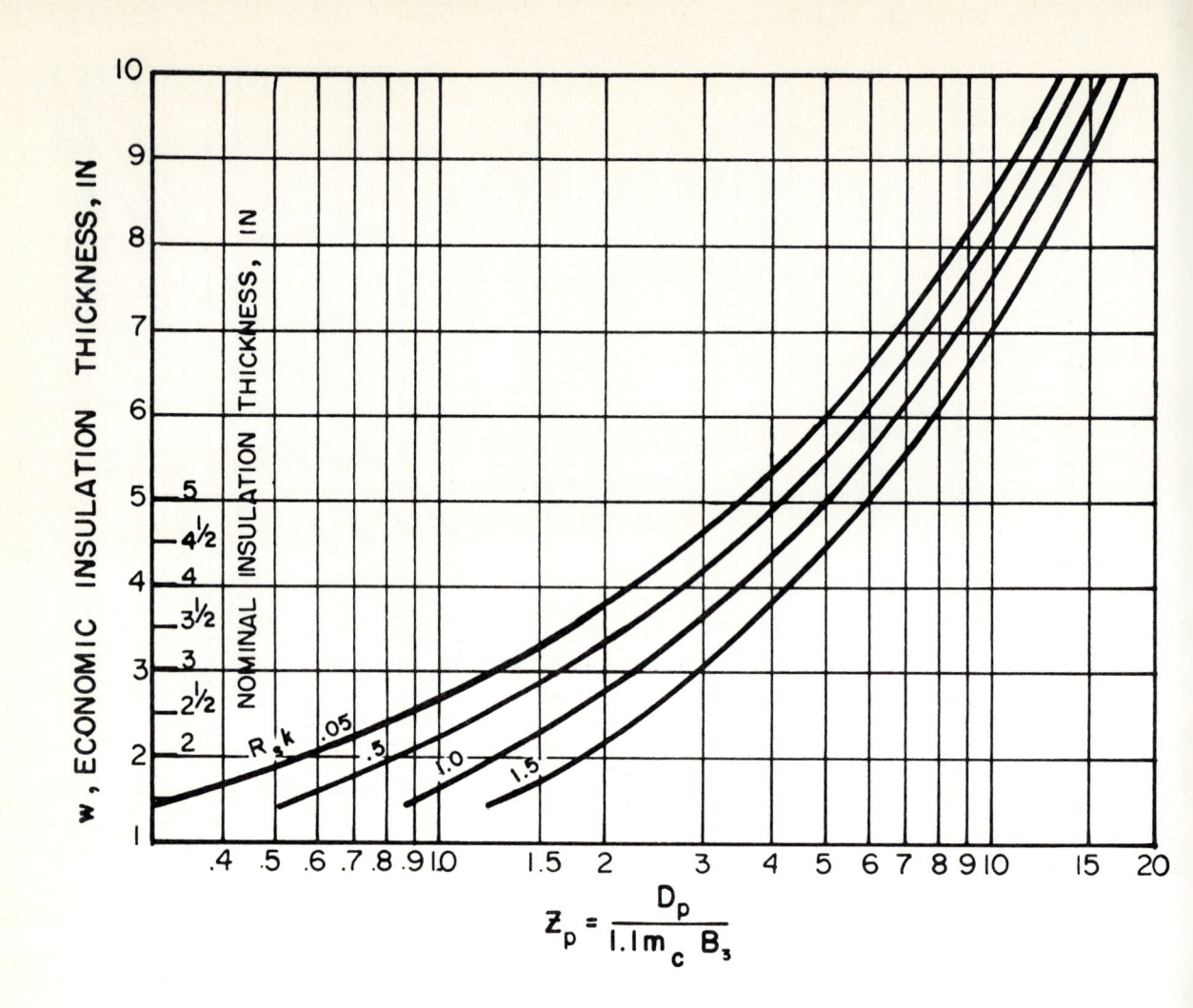

Figure 5-8o. w, Economic thickness, 14-inch pipe

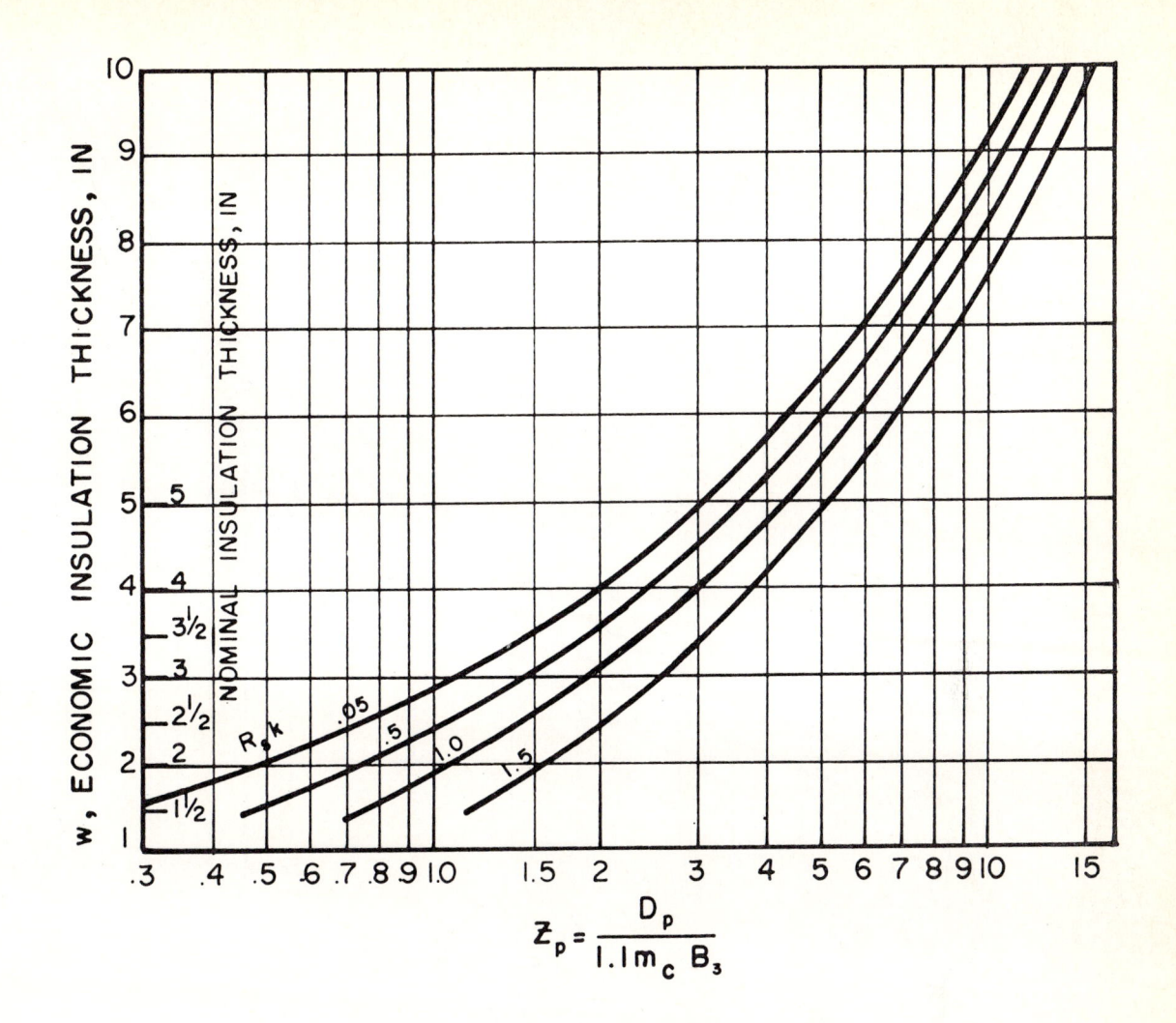

Figure 5-8p. w, Economic thickness, 16-inch pipe

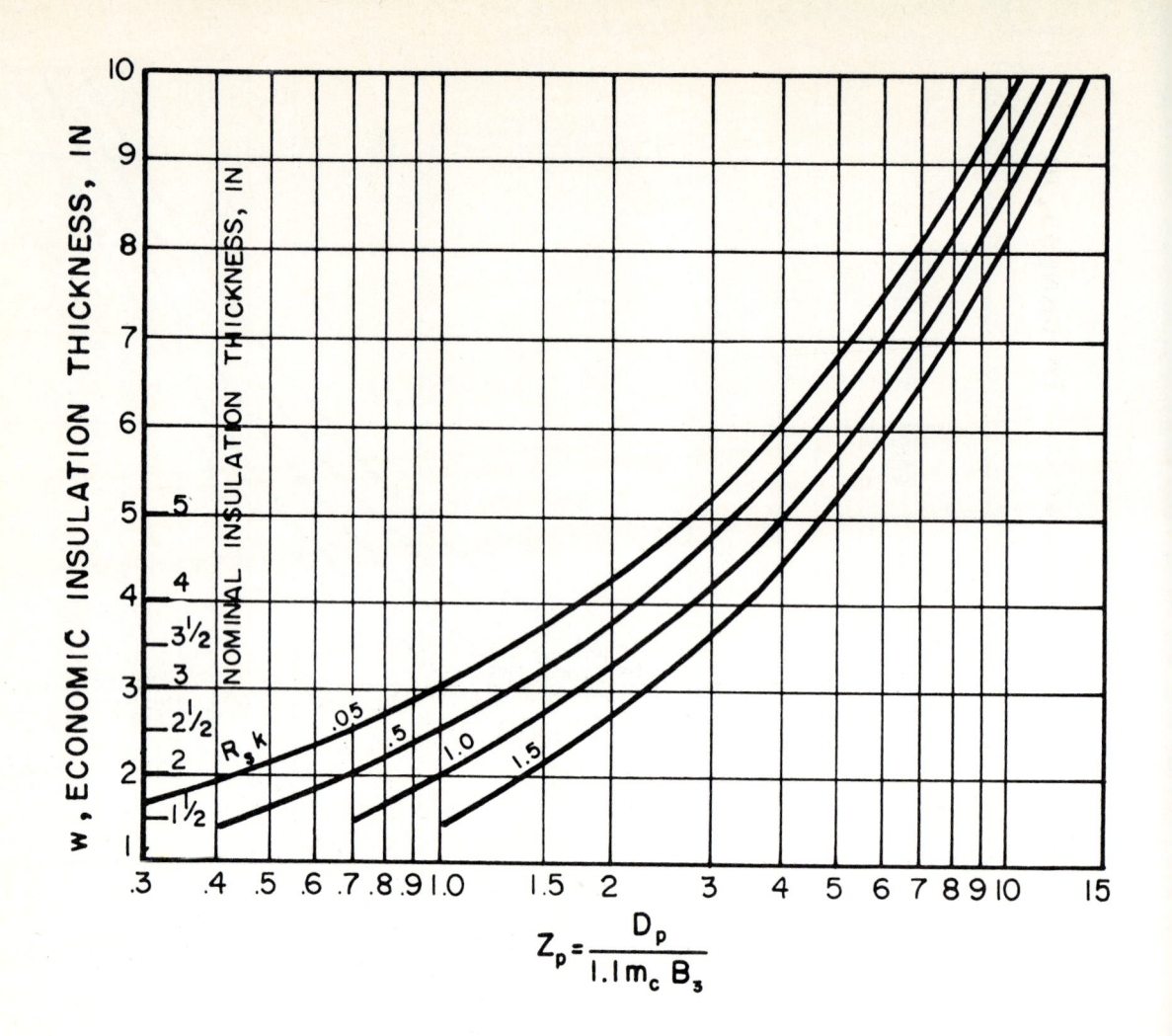

Figure 5-8q. w, Economic thickness, 18-inch pipe

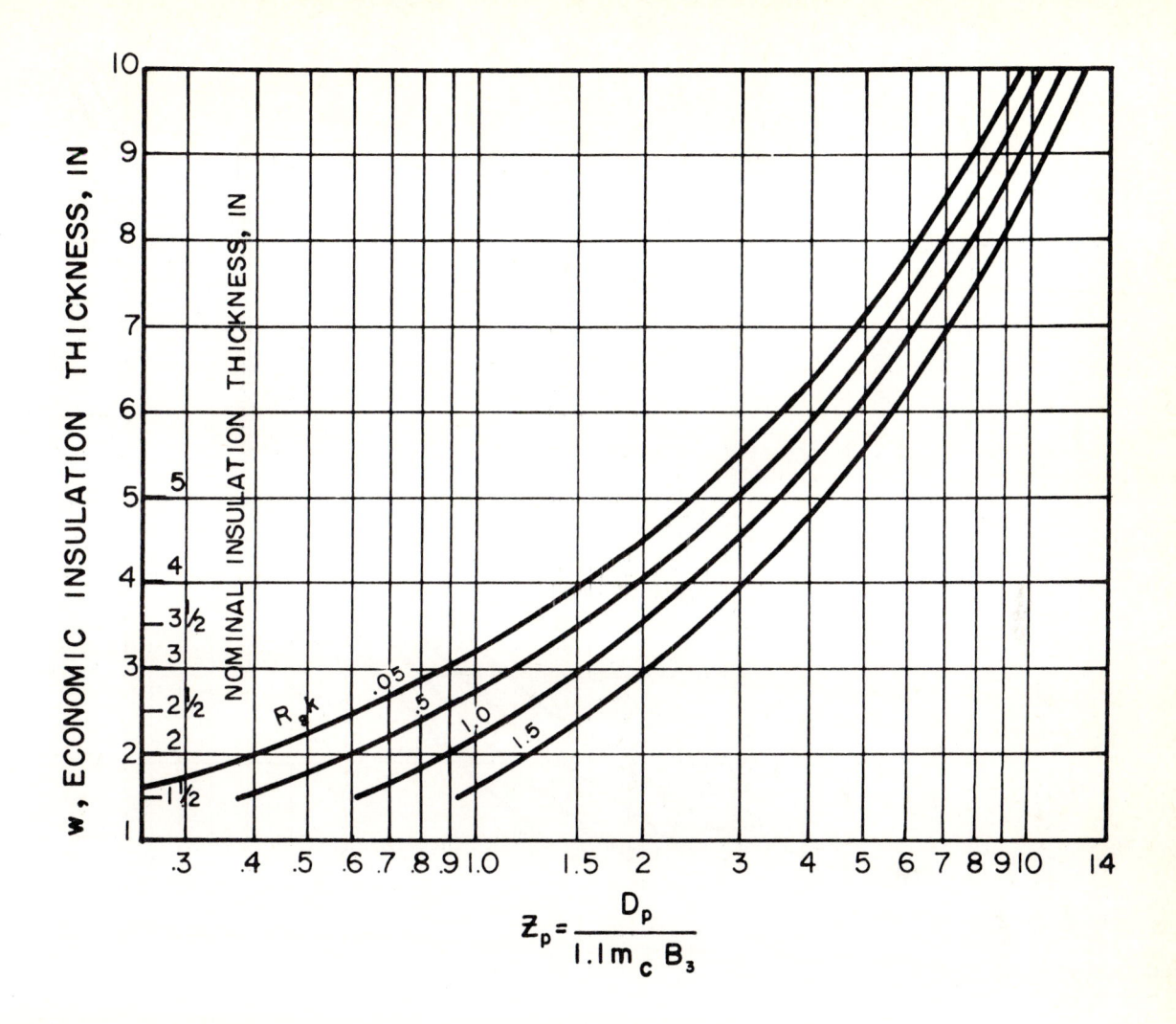

Figure 5-8r. w, Economic thickness, 20-inch pipe

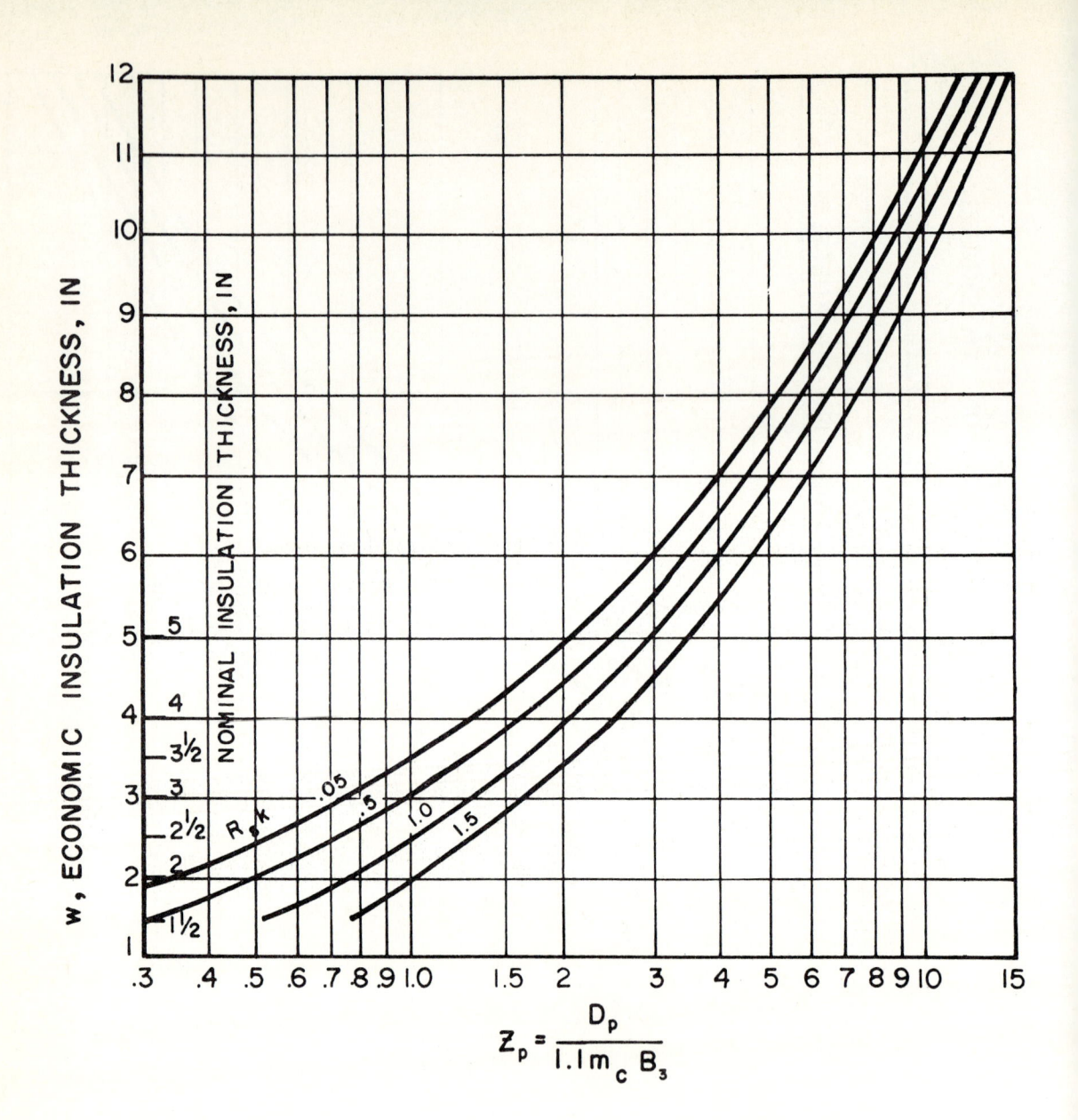

Figure 5-8s. w, Economic thickness, 24-inch pipe

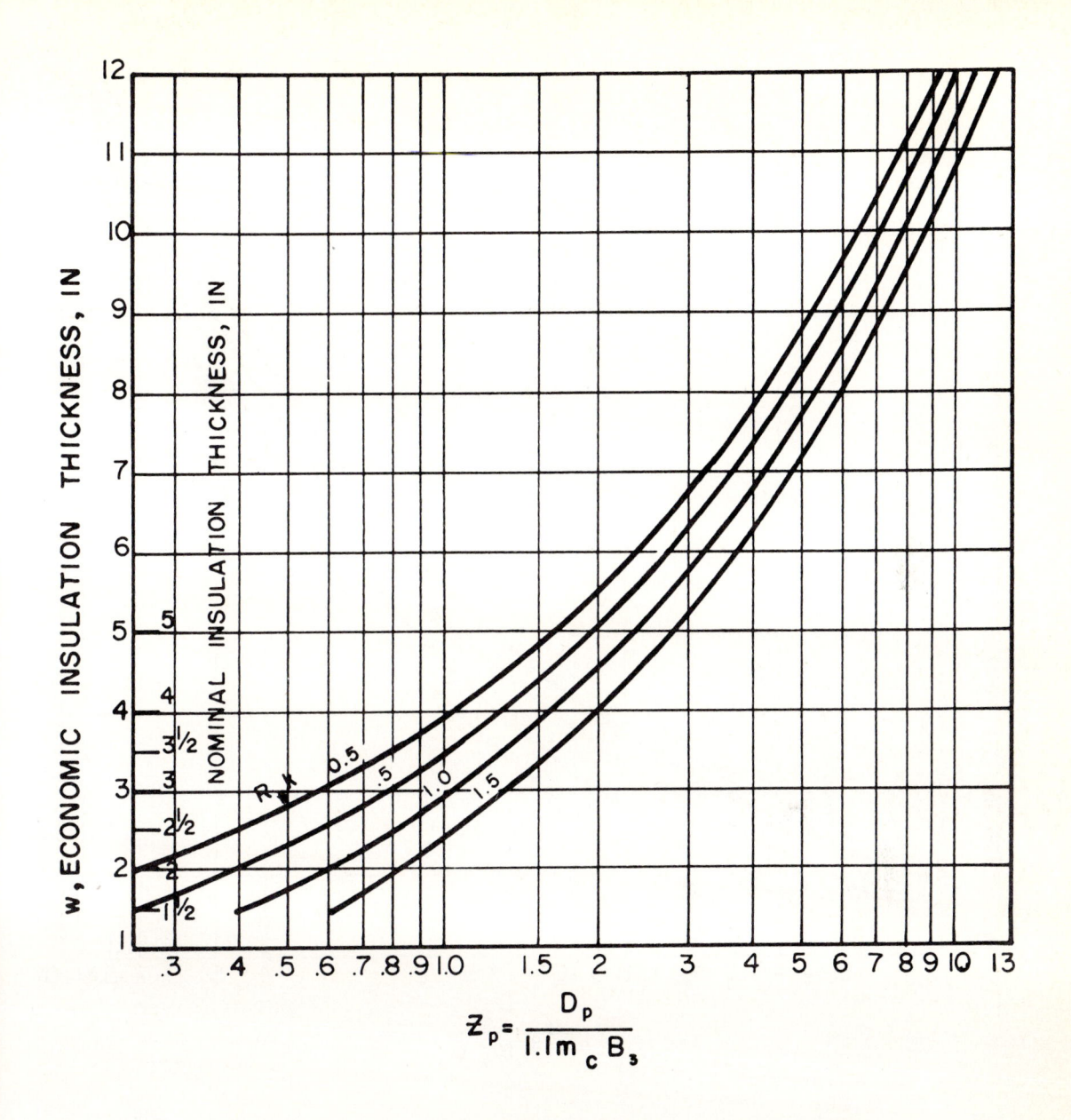

Figure 5-8t. w, Economic thickness, 30-inch pipe

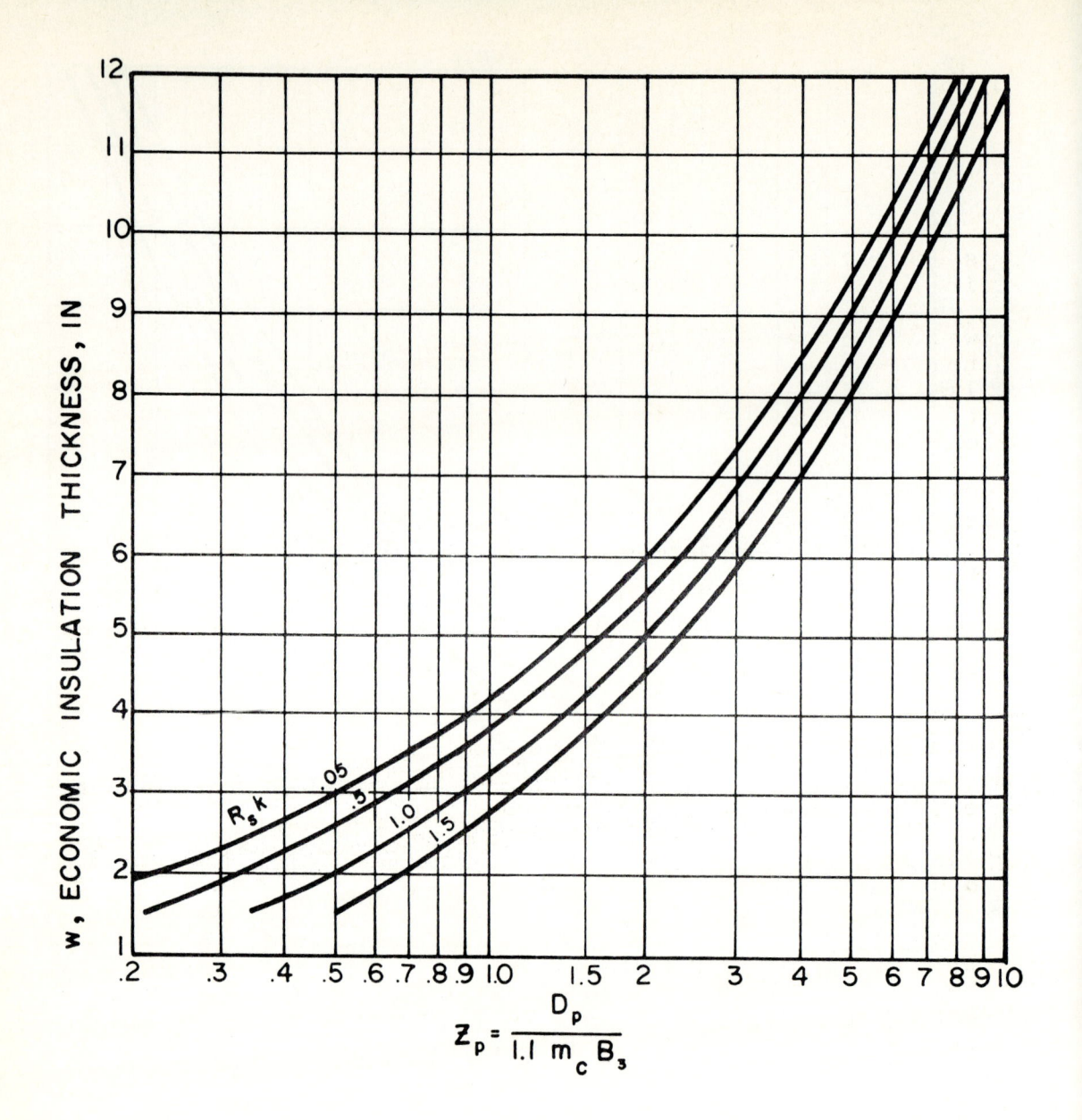

Figure 5-8u. w, Economic thickness, 36-inch pipe

CHAPTER 6
CONDENSATION CONTROL

Introduction

Pipes and equipment which operate below ambient temperatures require a certain minimum thickness of insulation to reduce heat gain and prevent surface condensation. When moisture is allowed to permeate an insulating material, its thermal effectiveness is reduced and eventual deterioration of the insulation results. Insulation for subambient temperature structures must therefore be specified in regard to both economics and condensation control.

This section is included for the purpose of insuring that the calculated economic thickness is sufficient to prevent condensation. A method for determining the thickness required to prevent condensation is presented.

The Condensation Problem

Moisture is transmitted from air to a cold surface by

1. condensation,
2. vapor pressure.

Condensation occurs when the temperature of the cold surface is below the dewpoint temperature of the air. The dewpoint temperature is that temperature at which air is saturated or incapable of holding additional water vapor.

The capacity of air to hold water vapor is a function of its temperature. The ratio of the quantity of water vapor in air to the maximum quantity that air is capable of containing at a specific temperature is termed relative humidity. Air is saturated when its relative humidity is 100 percent and the dewpoint temperature is equal to the ambient temperature.

The relative humidity of an air sample increases as its temperature decreases because the ability of air to hold water vapor is reduced. As air temperature is decreased, the saturation temperature may be reached, at which point condensation will occur on cold surfaces.

Moisture may also be transmitted from air to a cold surface by vapor pressure. As the temperature of a surface is lowered below that of the surrounding air, a lower vapor pressure area is created adjacent to the surface. Moisture is then forced to pass from the air to the surface to equalize the pressure. If the cold surface is insulated with material of open cell construction, water vapor will permeat the insulation, condense when the dewpoint temperature is reached, and thereby destroy the insulating value of that material. Wet insulation transmits heat at a rate approaching that of water which may be 50 to 100 times as great as that of the dry insulation. In systems where temperatures are low wet insulation may freeze, thereby causing the structural degradation or destruction of the insulation.

Although it is impossible to prevent condensation from forming under all ambient conditions, the following steps can be taken to minimize the problem and increase the useful life of the insulation.

1. Design the insulation system to resist moisture penetration by selecting a protective vapor or weather jacket and insuring that all jacket joints and seams are positively sealed.

2. Specify an insulation thickness that will maintain the surface temperature of the insulation above the dewpoint of the air.

Minimum thickness to prevent condensation is based upon total heat gain to the insulation surface from radiation and convection at the dewpoint temperature corresponding to design ambient air conditions. Each of the following criteria is considered in the calculation procedure:

1. Design ambient dry bulb temperature and relative humidity.
2. Insulation surface emissivity.
3. Process temperature.
4. Insulation thermal conductivity.

Determining Minimum Insulation Thickness

Glossary

ε	emmitance of the insulation finish, decimal unit
°F	degrees Fahrenheit
ft^2	square feet
hr	hour
in	inches
k	insulation conductivity, $\frac{Btu\text{-}in}{hr\text{-}ft^2\text{-}°F}$

L equivalent insulation thickness, in.

Q_a total heat flow from both radiation and convection, $\frac{Btu}{hr\text{-}ft^2}$

Q_{cv} heat flow from convection, $\frac{Btu}{hr\text{-}ft^2}$

Q_r heat flow from radiation, $\frac{Btu}{hr\text{-}ft^2}$

°R   degrees Rankine (460 + °F)

rh   relative humidity

t_a ambient dry bulb temperature, °F

t_d dewpoint temperature, °F

t_p temperature of pipe or surface to be insulated, °F

t_s temperature of the insulation finish surface, °F

t_m mean temperature of the insulation, °F

t_w wet bulb temperature, °F

w_c actual insulation thickness required to prevent condensation

Methodology

The ambient temperature and relative humidity conditions in which the cold equipment will be operating determine the temperature (dewpoint) at which condensation will form on cold surfaces. Therefore, the first design parameter to be established in specifying insulation for a cold system is the dewpoint temperature.

By using worst case dry bulb (t_a) and wet bulb (t_w) temperatures for the geographic area in question, the relative humidity can be determined and translated into a dewpoint temperature. The dewpoint temperature plus one degree (t_d + 1°F) is the design surface temperature (t_s).

In order to maintain minimum surface temperature, t_s, the amount of total heat flow into the insulation must be limited. The total allowable heat flow (Q_a) is found by summing the radiative (Q_r) and the convective (Q_{cv}) heat flows.

Radiation heat transferred to a body will depend on the surface emittance of that body. For instance, a dull insulation covering will emit a greater amount of radiative heat than will a shiny covering.

Knowing the design surface temperature (t_s), dry bulb temperature (t_a) and emittance (ε), the radiation heat flow (Q_r) into the insulation can be found. The convective heat flow (Q_{cv}) into the insulation is then found using ambient temperature (t_a) and design surface temperature (t_s). The sum of the heat gains from radiation and convection then becomes the total allowable heat gain ($Q_a = Q_r + Q_{cv}$).

The amount of insulation required to maintain the maximum allowable heat gain of Q_a under the design conditions can then be calculated, with thermal conductivity (k) of the insulation being an additional required consideration.

Condensation Control Worksheet

1. Ambient Design Air Conditions*

    a. Using Table 6-1 choose design dry bulb temperature. $t_a =$ _____, °F

    b. Using Table 6-1 choose design wet bulb temperature. $t_w =$ _____, °F

    c. Using Table 6-2 find design relative humidity. rh=0._____

    d. Using Fig. 6-1 find design dewpoint temperature. $t_d =$ _____, °F

    e. Determine design minimum surface temperature. $t_d + 1 = t_s$ $t_s =$ _____, °F

2. Allowable Heat Gain, Q_a

    a. Find surface emissivity (Table 6-3). $\varepsilon =$ _____

    b. Find radiation heat gain using Fig. 6-2. $Q_r =$ _____ Btu/hr-ft²

    c. Find convective heat gain using Fig. 6-3. $Q_{cv} =$ _____ Btu/hr-ft²

    d. Determine total allowable heat gain ($Q_a = Q_r + Q_{cv}$). $Q_a =$ _____ Btu/hr-ft²

3. Insulation Thickness

    a. Enter process temperature. $t_p =$ _____, °F

    b. Find insulation k value using Fig. 5-2 (Chapter 5). $k =$ _____ $\frac{\text{Btu-in}}{\text{hr-ft}^2\text{-°F}}$

    c. Find equivalent insulation thickness using Fig. 6-4. $L =$ _____, in

    d. Find actual pipe and flat surface thickness using Fig. 6-5. $w_c =$ _____, in

*Should the user have local data, that data should be used.

Table 6-1. Design Summer Conditions for the United States
(Temperature, °F)

Location	Design Dry Bulb 1%	2½%	5%	Design Wet Bulb 1%	2½%	5%	Location	Design Dry Bulb 1%	2½%	5%	Design Wet Bulb 1%	2½%	5%
ALABAMA							**CONNECTICUT**						
Anniston AP	96	94	93	79	78	77	Bridgeport AP	90	88	85	77	76	75
Auburn	98	96	95	80	79	78	Hartford, Brainard Field	90	88	85	77	76	74
Birmingham AP	97	94	93	79	78	77	New Haven AP	88	86	83	77	76	75
Dothan AP	97	95	94	81	80	70	New London	89	86	83	77	75	74
Gadsden	96	94	93	78	77	76	Norwich	88	86	83	77	76	75
Huntsville AP	97	95	94	78	77	76	Windsor Locks, Bradley Field	90	88	85	76	75	73
Mobile AP	85	93	91	80	79	79							
Montgomery AP	98	95	93	80	79	78	**DELAWARE**						
Selma-Craig AFB	98	96	94	81	80	79	Dover AFB	93	90	88	79	78	77
Tuscaloosa AP	98	96	95	81	80	79	Wilmington AP	93	90	87	79	77	76
ALASKA							**DISTRICT OF COLUMBIA**						
Anchorage AP	73	70	67	63	61	59	Andrews AFB	94	91	88	79	77	76
Barrow	58	54	50	54	51	48	Washington National AP	94	92	90	78	77	76
Fairbanks AP	82	78	75	64	63	61							
Juneau AP	75	71	68	66	64	62	**FLORIDA**						
Nome AP	66	62	59	58	56	54	Cape Kennedy AP	90	89	88	81	80	79
							Daytona Beach AP	94	92	91	81	80	79
ARIZONA							Fort Lauderdale	91	90	89	81	80	79
Douglas AP	100	98	96	70	69	68	Gainesville AP	96	94	93	80	79	79
Flagstaff AP	84	82	80	61	60	59	Jacksonville AP	96	94	92	80	79	79
Fort Huachuca AP	95	93	91	69	68	67	Key West AP	90	89	88	80	79	79
Nogales	100	98	96	72	71	70	Miami AP	92	90	89	80	79	79
Phoenix AP	108	106	104	77	76	75	Orlando AP	96	94	93	80	79	78
Prescott AP	86	94	91	67	66	65	Panama City, Tyndall AFB	92	91	90	81	80	80
Tucson AP	105	102	100	74	73	72	Pensacola Co	92	90	89	82	81	80
Yuma AP	111	109	107	79	78	77	St. Augustine	94	92	90	81	80	79
							St. Petersburg	93	91	90	81	80	79
ARKANSAS							Sarasota	93	91	90	80	80	79
El Dorado AP	98	96	95	81	80	79	Tallahassee AP	96	94	93	80	79	79
Fayetteville AP	97	95	93	77	76	75	Tampa AP	92	91	90	81	80	79
Hot Springs Nat. Pk.	99	97	96	79	78	77	West Palm Beach AP	92	91	89	81	80	79
Little Rock AP	99	96	94	80	79	78							
Texarkana AP	99	97	96	80	79	78	**GEORGIA**						
							Americus	98	96	93	80	79	78
CALIFORNIA							Atlanta AP	95	92	90	78	77	76
Bakersfield AP	103	101	99	72	71	70	Augusta AP	98	95	93	80	79	78
Blythe AP	111	109	106	78	77	76	Dalton	97	95	92	78	77	76
Burbank AP	97	94	91	72	70	69	Dublin	98	96	93	80	79	78
Crescent City AP	72	69	65	61	60	59	Gainesville	94	92	89	78	77	76
El Cajon	98	95	92	74	73	72	Macon AP	98	96	94	80	79	78
Escondido	95	92	89	73	72	71	Marietta, Dobbins AFB	95	93	94	78	77	76
Eureka/Arcata AP	67	65	63	60	59	58	Rome AP	97	95	93	78	77	76
Fairfield-Travis AFB	98	94	90	71	69	67	Savannah-Travis AP	96	94	92	81	80	79
Fresno AP	101	99	97	73	72	71							
Hamilton AFB	89	85	81	71	68	66	**HAWAII**						
Laguna Beach	83	80	77	69	68	67	Honolulu AP	87	85	84	75	74	73
Long Beach AP	87	84	81	72	70	69	Wahiawa	86	84	83	75	74	73
Los Angeles AP	86	83	80	69	68	67							
Merced-Castle AFB	102	99	96	73	72	70	**IDAHO**						
Modesto	101	98	96	72	71	70	Boise AP	96	93	91	68	66	65
Monterey	82	79	76	64	63	61	Idaho Falls AP	91	88	85	65	64	62
Napa	94	92	89	69	68	67	Lewiston AP	98	96	93	67	66	65
Needles AP	112	110	107	76	75	74	Pocatello AP	94	91	88	65	63	62
Oakland AP	85	81	77	65	63	62	Twin Falls AP	96	94	91	66	64	63
Oxnard AFB	84	80	78	70	69	67							
Palm Springs	110	108	105	79	78	77	**ILLINOIS**						
Pasadena	96	93	90	72	70	69	Aurora	93	91	88	78	77	75
Pomona CO	99	96	93	73	72	71	Bloomington	94	92	89	79	78	77
Redding AP	103	101	98	70	69	67	Chicago, O'Hare AP	93	90	87	77	75	74
Sacramento AP	100	97	94	72	70	69	Chicago CO	94	91	88	78	76	75
Salinas AP	87	85	82	67	65	64	Danville	96	94	91	79	78	76
San Diego AP	86	83	80	71	70	68	Galesburg	95	92	89	79	78	76
San Fernando AP	100	97	94	73	72	71	Greenville	96	94	92	79	78	77
San Francisco AP	83	79	75	65	63	62	Joliet AP	94	92	89	78	77	75
San Jose AP	90	88	85	69	67	65	La Salle/Peru	94	93	90	78	77	76
Santa Ana AP	92	89	86	72	71	70	Peoria AP	94	92	89	78	77	76
Santa Barbara CO	87	84	81	67	66	65	Pockford	92	90	87	77	76	75
Santa Cruz	87	84	80	66	65	63	Springfield AP	95	92	90	79	78	77
Santa Monica CO	80	77	74	69	68	67							
Santa Rosa	95	93	90	70	68	67	**INDIANA**						
Stockton AP	101	98	96	72	70	69	Bedford	95	93	90	79	78	77
Yuba City	102	100	97	71	70	69	Bloomington	95	92	90	79	78	76
							Evansville AP	96	94	91	79	78	77
COLORADO							Fort Wayne AP	93	91	88	77	76	75
Boulder	92	90	87	64	63	62	Huntington	94	92	89	78	76	75
Colorado Springs AP	90	88	86	63	62	61	Indianapolis AP	93	91	88	78	77	76
Denver AP	92	90	89	65	64	63	Jeffersonville	96	94	91	79	78	77
Greeley	94	92	89	65	64	63	Lafayette	94	92	89	78	77	76
La Junta AP	97	95	93	72	71	69	Marion	93	91	88	78	76	75
Pueblo AP	96	94	92	68	67	66	Muncie	93	91	88	78	77	75
Sterling	95	93	90	67	66	65	Richmond AP	93	91	88	78	77	75
							South Bend AP	92	89	87	77	76	74
							Terre Haute AP	95	93	91	79	78	77
							Valparaiso	92	90	87	78	76	75

Table 6-1. Design Summer Conditions for the United States, continued

Location	Design Dry Bulb 1%	2½%	5%	Design Wet Bulb 1%	2½%	5%	Location	Design Dry Bulb 1%	2½%	5%	Design Wet Bulb 1%	2½%	5%
IOWA							**MISSOURI**						
Burlington AP	95	92	89	80	78	77	Columbia AP	97	95	92	79	78	77
Cedar Rapids AP	92	90	87	78	76	75	Farmington AP	97	95	93	79	78	77
Des Moines AP	95	92	89	79	77	76	Hannibal	96	94	91	79	78	77
Dubuque	92	90	87	78	76	75	Jefferson City	97	95	93	79	78	77
Iowa City	94	91	88	79	77	76	Joplin AP	97	95	93	79	78	77
Marshalltown	93	91	88	79	77	76	Kansas City AP	100	97	94	79	77	76
Mason City AP	91	88	85	77	75	74	Mexico	96	94	91	79	78	77
Sioux City AP	96	93	90	79	77	76	St. Joseph AP	97	95	92	79	78	77
Waterloo	91	89	86	78	76	75	St. Louis AP	98	95	92	79	78	77
							Springfield AP	97	94	91	78	77	76
KANSAS													
Atchison	97	95	92	79	78	77	**MONTANA**						
Dodge City AP	99	97	95	74	73	72	Billings AP	94	91	88	68	66	65
El Dorado	101	99	96	78	77	76	Butte AP	86	83	80	60	59	57
Emporia	99	97	94	78	77	76	Glendive	96	93	90	71	69	68
Garden City AP	100	98	96	74	73	72	Great Falls AP	91	88	85	64	63	61
Great Bend	101	99	96	77	76	75	Helena AP	90	87	84	65	63	61
Hutchinson AP	101	99	96	77	76	75	Lewiston AP	89	86	83	65	63	62
Salina	101	99	96	78	76	75	Miles City	97	94	91	71	69	68
Topeka AP	99	96	94	79	78	77							
Wichita AP	102	99	96	77	76	75	**NEBRASKA**						
							Columbus	98	96	93	78	76	75
KENTUCKY							Fremont	99	97	94	78	77	76
Bowling Green AP	97	95	93	79	78	77	Hastings	98	96	94	77	75	74
Lexington AP	94	92	90	78	77	76	Lincoln CO	100	96	93	78	77	76
Louisville AP	96	93	91	79	78	77	Norfolk	97	95	92	78	76	75
Madisonville	96	94	92	79	78	77	Omaha AP	97	94	91	79	78	76
Paducah AP	97	95	94	80	79	78	Scottsbluff AP	96	94	91	70	69	67
LOUISIANA							**NEVADA**						
Baton Rouge AP	96	94	92	81	80	79	Carson City	93	91	88	62	61	60
Houma	94	92	91	81	80	79	Ely AP	90	88	86	60	59	58
Lafayette AP	95	93	92	81	81	80	Las Vegas AP	108	106	104	72	71	70
Lake Charles AP	95	93	91	80	79	79	Reno AP	95	92	90	64	62	61
Monroe AP	98	96	95	81	81	80	Tonopah AP	95	92	90	64	63	62
New Orleans AP	93	91	90	81	80	79							
							NEW HAMPSHIRE						
MAINE							Berlin	87	85	82	73	71	70
Augusta AP	88	86	83	74	73	71	Concord AP	91	88	85	75	73	72
Bangor, Dow AFB	88	86	81	75	73	71	Manchester, Grenier AFB	92	89	86	76	74	73
Lewiston	88	86	83	74	73	71	Portsmouth, Pease AFB	88	86	83	75	73	72
Portland AP	88	85	81	75	73	71							
							NEW JERSEY						
MARYLAND							Atlantic City CO	91	88	85	78	77	76
Baltimore AP	94	91	89	79	78	77	Long Branch	93	91	88	77	76	75
Cumberland	94	92	89	76	75	74	Newark AP	94	91	88	77	76	75
Hagerstown	94	92	89	77	76	75	Paterson	93	91	88	77	76	75
							Trenton CO	92	90	87	78	77	76
MASSACHUSETTS													
Boston AP	91	88	85	76	74	73	**NEW MEXICO**						
Fall River	88	86	83	75	74	73	Albuquerque AP	96	94	92	66	65	64
Framingham	91	89	86	76	74	73	Clovis AP	99	97	95	70	69	68
Gloucester	86	84	81	74	73	72	Farmington AP	95	93	91	66	65	64
New Bedford	86	84	81	75	73	72	Los Alamos	88	86	83	64	63	62
Pittsfield AP	86	84	81	74	72	71	Santa Fe CO	90	88	85	65	63	62
Springfield, Westover AFB	91	88	85	76	74	73	Socorro AP	99	97	94	67	66	65
Worcester AP	89	87	84	75	73	71							
							NEW YORK						
MICHIGAN							Albany AP	91	88	85	76	74	73
Battle Creek AP	92	89	86	76	74	73	Buffalo AP	88	86	83	75	73	72
Detroit Met. CAP	92	88	85	76	75	74	Cortland	90	88	85	75	73	72
Flint AP	89	87	84	76	75	74	Elmira AP	92	90	87	75	73	72
Grand Rapids AP	91	89	86	76	74	73	Ithaca	91	88	85	75	73	72
Jackson AP	92	80	86	76	75	74	Kingston	92	90	87	76	74	73
Kalamazoo	92	89	86	76	75	74	Newburgh-Stewart AFB	92	89	86	78	76	74
Lansing AP	89	87	84	76	75	73	NYC-Central Park	94	91	88	77	76	75
Marquette CO	88	86	83	73	71	69	NYC-LaGuardia AP	93	90	87	77	76	75
Muskegon AP	87	85	82	75	74	73	Niagara Falls AP	88	86	83	75	74	73
Pontiac	90	88	85	76	75	73	Oswego CO	86	84	81	75	74	72
Port Huron	90	88	85	76	74	73	Plattsburg AFB	86	84	81	74	73	71
Saginaw AP	88	86	83	76	75	73	Poughkeepsie	93	90	87	77	75	74
Traverse City AP	89	86	83	75	73	72	Rome-Griffis AFB	90	87	84	76	74	73
							Schenectady	90	88	85	75	73	72
MINNESOTA							Syracuse AP	90	87	85	76	74	73
Duluth AP	85	82	79	73	71	69	Utica	89	87	84	75	73	72
Faribault	90	88	85	77	75	74							
International Falls AP	86	82	79	72	69	68	**NORTH CAROLINA**						
Minneapolis/St.Paul AP	92	89	86	77	75	74	Charlotte AP	96	94	92	78	77	76
Rochester AP	90	88	85	77	75	74	Durham	94	92	89	78	77	76
St. Cloud AP	90	88	85	77	75	73	Fayetteville, Pope AFB	97	94	92	80	79	78
							Greensboro AP	94	91	89	77	76	75
MISSISSIPPI							Henderson	94	92	89	79	78	77
Columbus AFB	97	95	93	79	79	78	Jacksonville	94	92	89	81	80	79
Greenwood	98	96	94	81	80	79	Rocky Mount	95	93	90	80	79	78
Jackson AP	98	96	94	79	78	78	Winston-Salem AP	94	91	89	77	76	75
Laurel	97	95	94	80	79	78							
Meridian AP	97	95	94	80	79	78							
Tupelo	98	96	95	80	79	78							

Table 6-1. Design Summer Conditions for the United States, continued

Location	Design Dry Bulb 1%	2½%	5%	Design Wet Bulb 1%	2½%	5%	Location	Design Dry Bulb 1%	2½%	5%	Design Wet Bulb 1%	2½%	5%
NORTH DAKOTA							TEXAS						
Bismarck AP	95	91	88	74	72	70	Abilene AP	101	99	97	76	75	74
Grand Forks AP	91	87	84	74	72	70	Amarillo AP	98	96	93	72	71	70
Jamestown AP	95	91	88	75	73	71	Austin AP	101	98	96	79	78	77
Minot AP	91	88	84	72	70	68	Beaumont	96	94	93	81	80	79
							Brownsville AP	94	92	91	80	80	79
OHIO							Corpus Christi AP	95	93	91	81	80	80
Akron/Canton AP	89	87	84	75	73	72	Dallas AP	101	99	97	79	78	78
Cincinnati CO	94	92	90	78	77	76	El Paso AP	100	98	96	70	69	68
Cleveland AP	91	89	86	76	75	74	Fort Worth AP	102	100	98	79	78	77
Columbus AP	92	88	86	77	76	75	Galveston AP	91	89	88	82	81	81
Dayton AP	92	90	87	77	75	74	Houston	96	94	92	80	80	79
Fremont	92	90	87	76	75	74	Laredo AFB	103	101	100	79	78	78
Lancaster	93	91	88	77	76	75	Midland AP	100	98	96	74	73	72
Marion	93	91	88	77	76	75	Pampa	100	98	95	73	72	71
Middletown	93	91	88	77	76	75	Pecos	102	100	97	72	71	70
Portsmouth	94	92	89	77	76	75	Port Arthur AP	94	92	91	81	80	80
Springfield	93	90	88	77	76	75	San Antonio AP	99	97	96	77	77	76
Toledo AP	92	90	87	77	75	74	Waco AP	101	99	98	79	78	78
Warren	90	88	85	75	74	73	Wichita Falls AP	103	100	98	77	76	75
Youngstown AP	89	86	84	75	74	73							
							UTAH						
OKLAHOMA							Richfield	94	92	89	66	65	64
Altus AFB	103	101	99	77	76	75	St. George CO	104	102	99	71	70	69
Bartlesville	101	99	97	79	78	77	Salt Lake City AP	97	94	92	67	66	65
Lawton AP	103	101	98	78	77	76							
Muskogee AP	102	99	96	79	78	77	VERMONT						
Oklahoma City AP	100	97	95	78	77	76	Barre	86	84	81	73	72	70
Stillwater	101	99	97	78	77	76	Burlington AP	88	85	83	74	73	71
Tulsa AP	102	99	96	79	78	77	Rutland	87	85	82	74	73	71
OREGON							VIRGINIA						
Albany	91	88	84	69	67	65	Charlottsville	93	90	88	79	77	76
Baker AP	94	92	89	66	65	63	Danville AP	95	92	90	78	77	76
Eugene AP	91	88	84	69	67	65	Fredricksburg	94	92	89	79	78	76
Klamath Falls AP	89	87	84	63	62	61	Harrisonburg	92	90	87	78	77	76
Pendleton AP	97	94	91	66	65	63	Lynchburg AP	94	92	89	77	76	75
Portland AP	89	85	81	69	67	66	Norfolk AP	94	91	89	79	78	78
Salem AP	92	88	84	69	67	66	Richmond AP	96	93	91	79	78	77
							Roanoke AP	94	91	89	76	75	74
PENNSYLVANIA							Winchester	94	92	89	78	76	75
Allentown AP	92	90	87	77	75	74							
Butler	91	89	86	75	74	73	WASHINGTON						
Erie AP	88	85	82	76	74	73	Aberdeen	83	80	77	62	61	60
Harrisburg AP	92	89	86	76	75	74	Olympia AP	85	83	80	67	65	63
Lancaster	92	90	87	77	76	75	Seattle-Boeing Fld	82	80	77	67	65	64
Philadelphia	93	90	87	78	77	76	Seattle CO	81	79	76	67	65	64
Pittsburgh AP	90	87	85	75	74	73	Seattle-Tacoma AP	85	81	77	66	64	63
Reading CO	92	90	87	77	76	75	Spokane AP	93	90	87	66	64	63
State College	89	87	84	74	73	72	Tacoma-McChord AFB	85	81	78	68	66	64
Uniontown	90	88	85	75	74	73	Wenatchee	95	92	89	68	66	64
West Chester	92	90	87	77	76	75							
York	93	91	88	77	76	75	WEST VIRGINIA						
							Bluefield AP	88	86	83	74	73	72
RHODE ISLAND							Charleston AP	92	90	88	76	75	74
Newport	86	84	81	75	74	73	Clarksburg	92	90	87	76	75	74
Providence AP	89	86	83	76	75	74	Huntington CO	95	93	91	77	76	75
							Wheeling	91	89	86	76	75	74
SOUTH CAROLINA													
Charleston AFB	94	92	90	81	80	79	WISCONSIN						
Charleston CO	95	93	90	81	80	79	Appleton	89	87	84	75	74	72
Columbia AP	98	96	94	79	79	78	Eau Claire AP	90	88	85	76	74	72
Greenwood	97	95	92	78	77	76	Green Bay AP	88	85	82	75	73	72
Rock Hill	97	95	92	78	77	76	Madison AP	92	88	85	77	75	73
Spartanburg AP	95	93	90	77	76	75	Milwaukee AP	90	87	84	77	75	73
							Waukesha	91	90	86	77	75	74
SOUTH DAKOTA							Wausau AP	89	86	83	74	72	70
Aberdeen AP	95	92	89	77	75	74							
Mitchell	96	94	91	77	76	74	WYOMING						
Rapid City AP	96	94	91	72	71	69	Casper AP	92	90	87	63	62	60
Sioux Falls AP	95	92	89	77	75	74	Cheyenne AP	89	86	83	63	62	61
Yankton	96	94	91	78	76	75	Laramie AP	82	80	77	61	59	58
							Newcastle	92	89	86	68	67	66
TENNESSEE							Sheridan AP	95	92	89	67	65	64
Chattanooga AP	97	94	92	78	78	77							
Clarksville	98	96	94	79	78	77							
Columbia	97	95	93	79	78	77							
Greenville	93	91	88	76	75	74							
Jackson AP	97	95	94	80	79	78							
Knoxville AP	95	92	90	77	76	75							
Memphis AP	98	96	94	80	79	78							
Nashville AP	97	95	92	79	78	77							

Note.—1%, 2½% and 5% represent the number of hours during summer months of June through September (2,928 hours) in which these design temperatures are surpassed.

Adjustment for elevation: For a location at a lower elevation the design values should be increased; at higher elevations they should be decreased.

Dry bulb temperature, 1°F per 200 ft.
Wet bulb temperature, 1°F per 500 ft.

Source: ASHRAE Handbook of Fundamentals.

Table 6-2. Table of Relative Humidities, in Percent

Dry Bulb Temp.	\multicolumn{24}{c}{Difference Between Readings of Wet and Dry Bulbs, °F}
	1	2	3	4	5	6	7	8	9	10	11	12	13	14	15	16	17	18	19	20	22	24	26	28	30	32	34	36	38	40	45
30	89	78	67	56	46	36	26	16	6	0																					
35	91	81	72	63	54	45	36	27	19	10	2	0																			
40	92	83	75	68	60	52	45	37	29	22	15	7	0																		
45	93	86	78	71	64	57	51	44	38	31	25	18	12	6	0																
50	93	87	80	74	67	61	55	49	43	38	32	27	21	16	10	5	0														
55	94	88	82	76	70	65	59	54	49	43	38	33	28	23	19	14	9	5	0	0											
60	94	88	83	78	73	68	63	58	53	48	43	39	34	30	26	21	17	13	9	5	0										
65	95	90	85	80	75	70	66	61	56	52	48	44	39	35	31	27	24	20	16	12	5	0									
70	95	90	86	81	77	72	68	64	59	55	51	48	44	40	36	33	29	25	22	19	12	6	0								
75	96	91	86	82	78	74	70	66	62	58	54	51	47	44	40	37	34	30	27	24	18	12	7	1	0						
80	96	91	87	83	79	75	72	68	64	61	57	54	50	47	44	41	38	35	32	29	23	18	12	7	3	0					
85	96	92	88	84	80	76	73	70	66	63	59	56	53	50	47	44	41	38	35	32	27	22	17	13	4	0					
90	96	92	89	85	81	78	74	71	68	65	61	58	55	52	49	47	44	41	39	36	31	26	22	17	13	9	5	1	0		
95	96	92	89	85	82	79	75	72	69	66	63	60	57	54	52	49	46	43	42	38	34	30	25	21	17	13	9	6	2	0	
100	96	93	89	86	83	80	77	74	71	69	65	62	59	57	54	52	49	47	45	43	38	34	30	26	23	20	16	13	10	6	
102	96	93	89	86	83	80	77	74	71	69	65	62	59	57	54	52	49	47	45	43	38	34	30	26	23	20	16	13	10	6	
104	96	93	90	86	83	80	77	74	71	65	65	63	60	58	55	52	50	48	46	43	39	35	31	27	23	21	18	15	12	8	1
106	96	93	90	87	83	80	77	74	69	66	66	63	60	58	56	53	51	48	46	55	40	36	32	28	24	21	19	16	13	10	1
108	96	93	90	87	84	81	78	75	72	70	66	64	61	59	56	54	51	49	47	45	41	37	33	29	26	22	19	17	14	11	5
110	96	93	90	87	84	81	78	75	72	70	67	64	62	60	57	55	52	50	48	46	41	37	34	30	27	23	20	17	15	12	6

89

Table 6-3. Emittance Ranges for Metal Jackets, Mastics, and Various Surface Finishes

Type of Covering	Condition	Emittance ε (at surface temperature of approx. 100°F)
Aluminum	Polished	0.03 to 0.06
	Gray, dull	0.06 to 0.09
	Oxidized	0.10 to 0.20
Aluminum paint	New	0.20 to 0.30
	After weathering	0.40 to 0.70
Asbestos paper	Clean	0.90 to 0.94
Asphalt asbestos felts	do.	0.93 to 0.96
Asphalt mastics	do.	0.90 to 0.95
Galvanized steel	New, bright	0.06 to 0.10
	Dull	0.20 to 0.30
Paints	White, clean	0.55 to 0.70
	Green, clean	0.65 to 0.80
	Gray, clean	0.80 to 0.90
	Black, clean	0.90 to 0.95
Painted canvas	Color as painted	Will be approximately the same as ε for color of paint used.
PVA and PVC mastics	White, clean	0.60 to 0.70
	Green, clean	0.70 to 0.80
	Gray, medium-clean	0.85 to 0.90
	Black	0.85-0.95
Stainless steel	Polished	0.22-0.26
	No. 4 mill finish	0.35 to 0.40
	Oxidized	0.80 to 0.85

Source: Malloy, J. F., Thermal Insulation, (Van Nostrand Reinhold Co., New York), 1969, Table 7-41, p. 527.

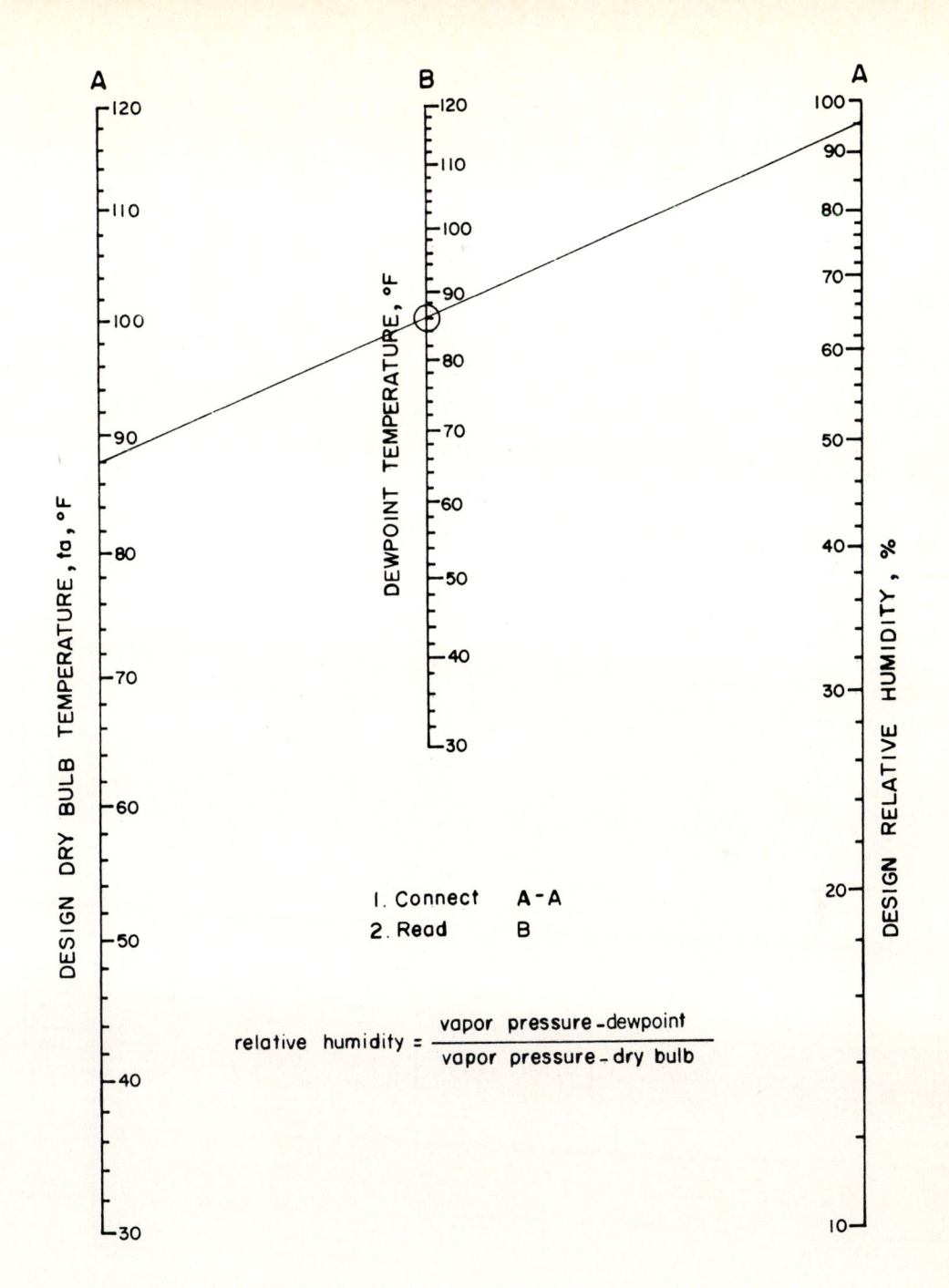

Figure 6-1. Dewpoint from relative humidity and dry bulb temperature

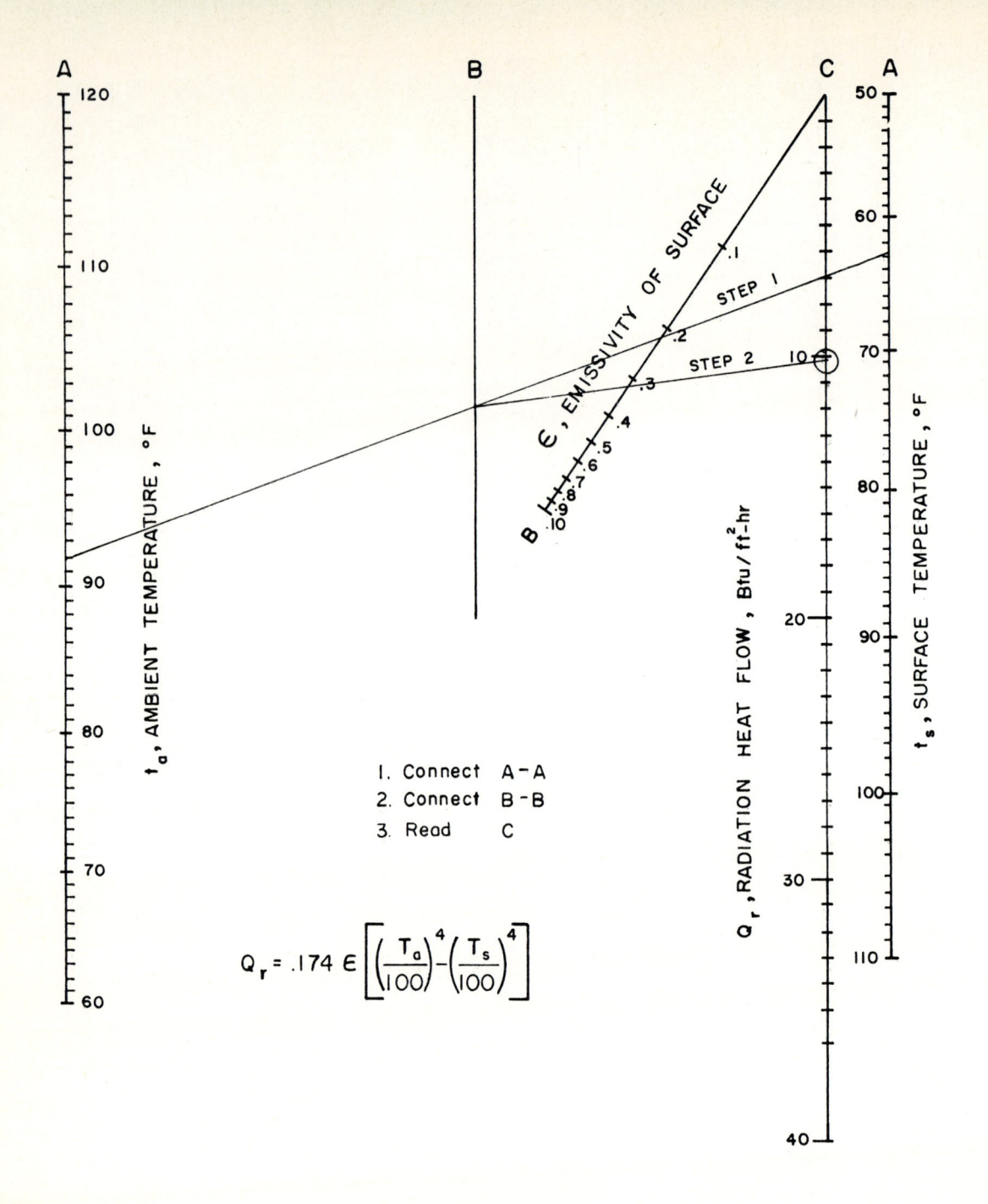

RADIATION HEAT FLOW

Figure 6-2. Radiation heat flow from surface of insulation

$$Q_r = .174\, \epsilon \left[\left(\frac{T_a}{100}\right)^4 - \left(\frac{T_s}{100}\right)^4\right]$$

1. Connect A-A
2. Connect B-B
3. Read C

Figure 6-3. Convection heat flow from surface of insulation

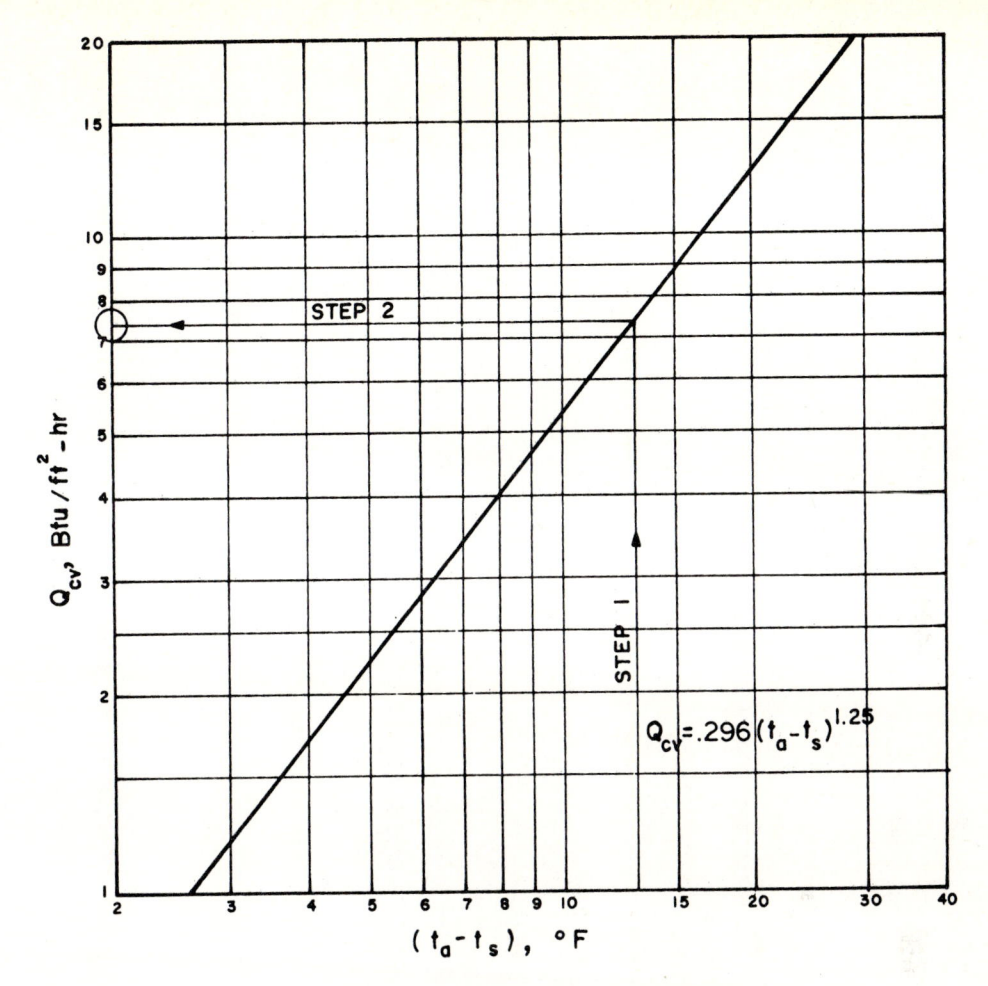

CONVECTION HEAT FLOW

$Q_{cv} = .296(t_a - t_s)^{1.25}$

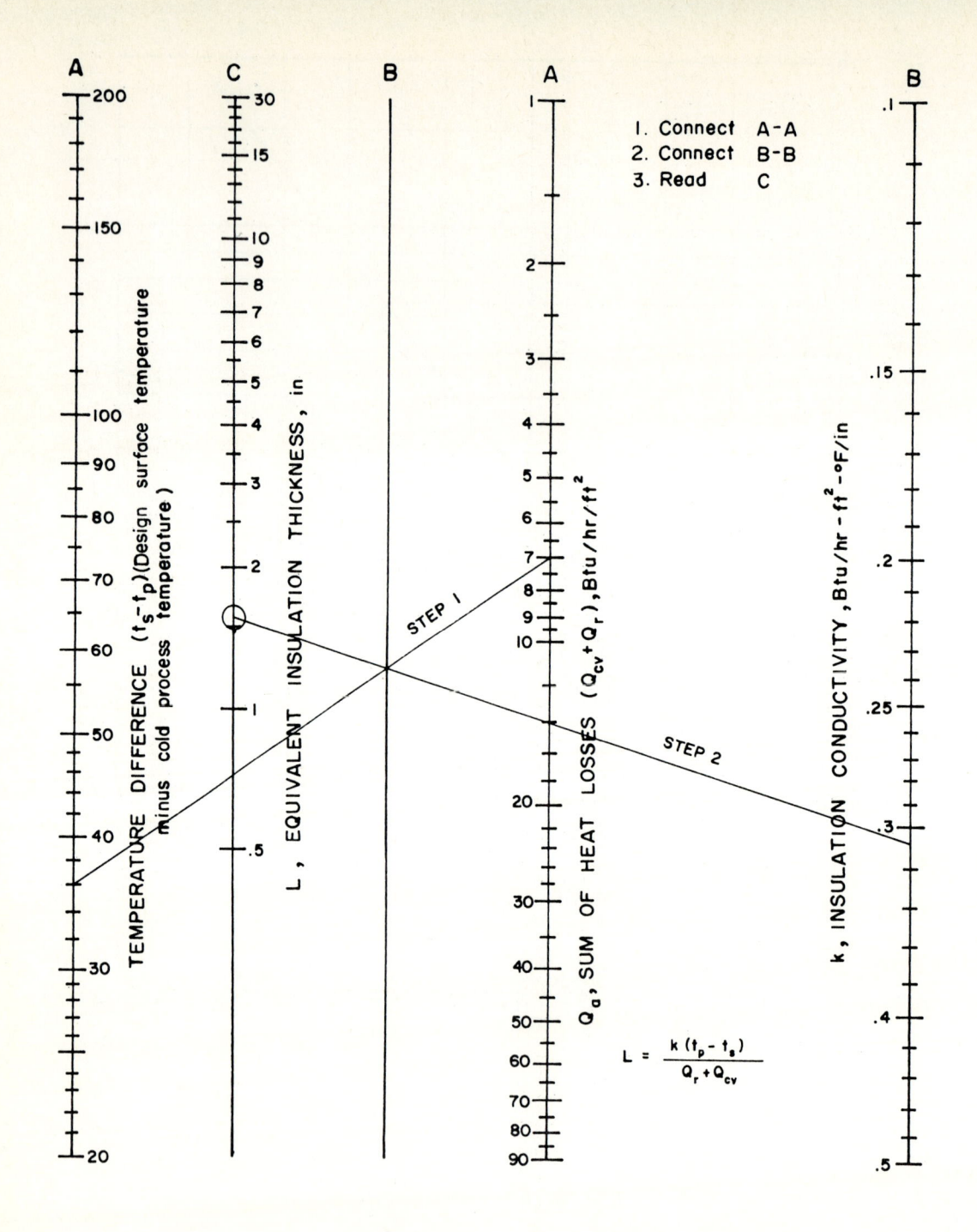

EQUIVALENT INSULATION THICKNESS REQUIRED
TO PREVENT CONDENSATION

Figure 6-4. Insulation thickness for condensation prevention

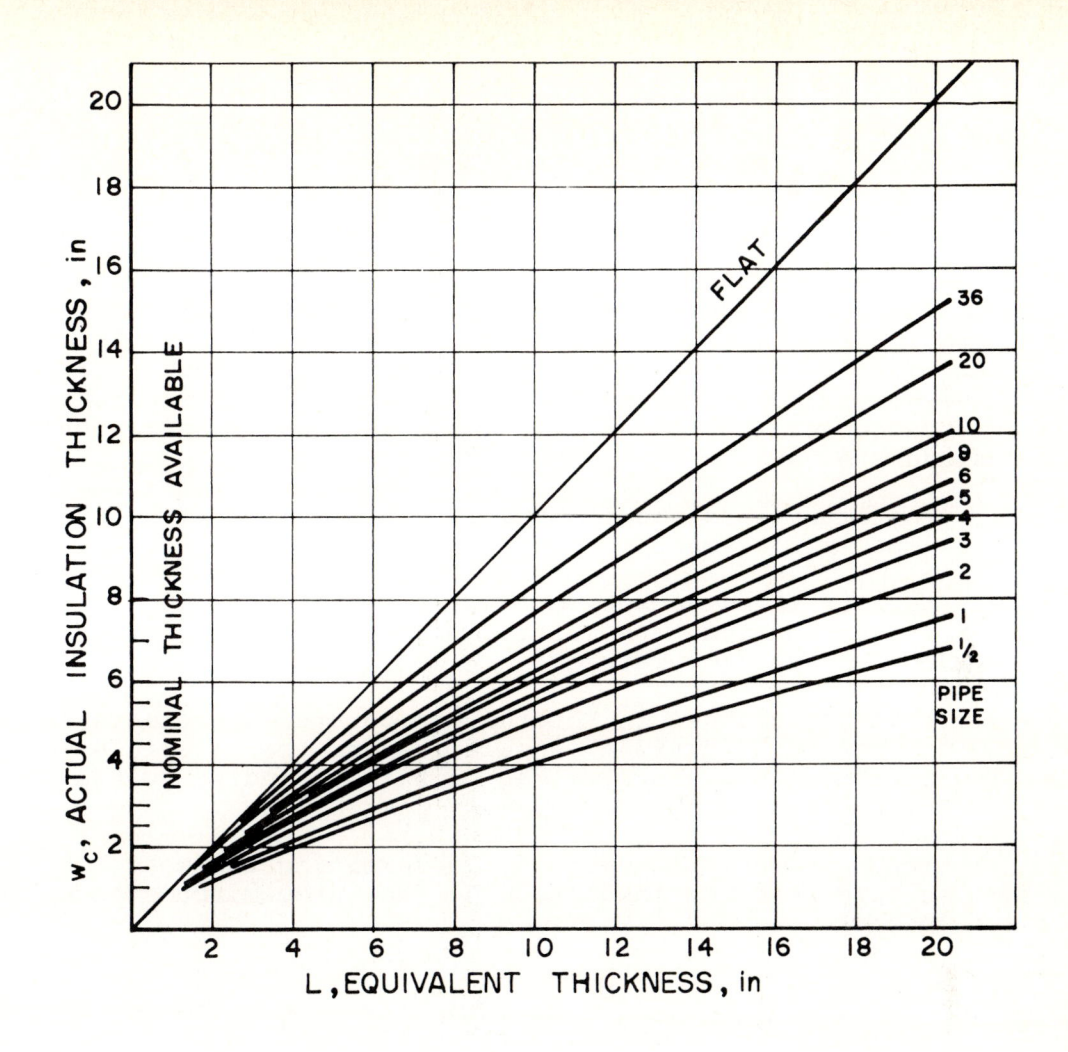

Figure 6-5. Actual insulation thickness required to prevent condensation

CHAPTER 7
RETROFITTING INSULATION

Introduction

Retrofitting is defined as the application of additional insulation over existing insulation, new insulation after old insulation has been removed, or new insulation over existing, previously uninsulated surfaces. Retrofitting causes a change in the interface temperature between the added insulation and the existing surface. The designer is cautioned to make certain that:

1. The dimensions are compatable (in the case of pipe insulations)

2. That existing jackets or environmental coverings, if not removed, are suitable for continued service at the new temperature.

3. That the changes in internal temperature gradients across the additional insulation and the materials in the existing system can be safely handled by the components of the new system.

This section provides procedures for determining the most economical thickness for insulation in the three circumstances defined above.

Procedure

Case 1, new insulation over existing insulation. A more precise solution for this case is presented in Appendix E. For an adequate solution the following simplified procedure is recommended.

Glossary

B	annual amortization factor to amortize cost of insulation
Btu	British thermal unit
D_p	annual cost of heat lost through one linear foot of pipe insulation
D_s	annual cost of heat lost through one ft^2 flat surface insulation
°F	degrees Fahrenheit
ft	feet

i_3 cost of insulation rate or required rate of return on the last increment of insulation

in   inch

k_1 coefficient of thermal conductivity for existing insulation, $\dfrac{\text{Btu-in}}{\text{hr-ft}^2\text{-}°F}$

k_2 coefficient of thermal conductivity for existing insulation, $\dfrac{\text{Btu-in}}{\text{hr-ft}^2\text{-}°F}$

k_m average coefficient of thermal conductivity, $\dfrac{\text{Btu-in}}{\text{hr-ft}^2\text{-}°F}$

M project average annual cost of heat, $\dfrac{\$}{10^6 \text{Btu}}$

m_c insulation incremental cost, $\dfrac{\$}{\text{in-lin ft}}$

n_1 term of insulation project, yr

R_s surface resistivity of the insulation system, $\dfrac{\text{hr-ft}^2\text{-}°F}{\text{Btu}}$

t_a temperature of ambient dry bulb, $°F$

t_{m1} mean temperature of existing insulation, $°F$

t_{m2} mean temperature of new insulation, $°F$

t_p temperature of pipe or surface to be insulated, $°F$

ΔT $t_p - t_a$, $°F$

w economic thickness of insulation, inches

w_t economic thickness to add to existing insulaton, in

Y annual hours of operation, hr

Z factor used in economic thickness determination

$\$$ dollars

Retrofit Insulation Worksheet

Case 1-Adding Over Existing Insulation

1. Enter process temperature, t_p. _____ ,°F

2. Enter ambient temperature, t_a. _____ ,°F

3. Calculate mean temperature for existing insulation

   $$t_{m1} = \frac{3t_p + t_a}{4}$$ _____ ,°F

4. Find thermal conductivity for t_{m1} using Fig. 5-1 or 5-2. $k_1 =$ _____ $\frac{\text{Btu-in}}{\text{hr-ft}^2\text{-°F}}$

5. Calculate mean temperature for new insulation layer

   $$t_{m2} = \frac{t_p}{4} + \frac{3 t_a}{4}$$ $t_{m2} =$ _____ ,°F

6. Find thermal conductivity for t_{m2} using Fig. 5-1 or 5-2. $k_2 =$ _____ $\frac{\text{Btu-in}}{\text{hr-ft}^2\text{-°F}}$

7. Calculate average thermal conductivity:

   $$\frac{k_1 + k_2}{2}$$ $k_m =$ _____ $\frac{\text{Btu-in}}{\text{hr-ft}^2\text{-°F}}$

8. Calculate $\Delta T = t_p - t_a$. $\Delta T =$ _____ ,°F

9. Find cost of heat, M, using Chapter 3. $M =$ _____ $/10^6$ Btu

10. Find insulation incremental cost, m_c, from Chapter 4. Use the pipe size nearest in outer diameter to that of the existing insulation. $m_c =$ _____ $/in-1in-ft

11. Enter annual hours of operation, Y. $Y =$ _____ hrs

12. Find D for flat surface or D_p for pipe using Figure 5-3. $D_s =$ _____

    $D_p =$ _____

13. Determine B using Fig. 5-4 and Fig. 5-5. Use i_3, money cost for insulation and n_1, insulation project life.

i_3 = 0._____

n_1 = _____ yrs

B = _____

14. Find Z_s using Fig. 5-6 for flat, or Z_p using Fig. 5-7 for pipe.

Z_s = _____

Z_p = _____

15. Calculate $k_m R_s$.

    (R_s of 0.7 is typical.)

$k_m R_s$ = _____

16. Use proper Fig. 5-8 to determine economic thickness. (Use pipe size, not existing insulation diameter.)

w = _____ inches

17. Subtract existing insulation thickness from w to find economic retrofit thickness to add $w_t = w-(r_2-r_1)$.

w_t = _____ inches

Case 2-New Insulation Over Bare Pipe After Stripping Old Insulation

Proceed as in new work, see Chapters 3, 4, 5, 6.

Case 3-New Insulation Over Previously Uninsulated Surfaces

Proceed as in new work, but make allowances for installation difficulties, such as lack of space, when estimating m_c in Chapter 4. See Chapters 3, 4, 5, 6.

CHAPTER 8
SAMPLE PROBLEMS

No. 1—High-Temperature Steam Pipe

A new electric utility installation will have a 1,000-foot long, 16-inch, NPS high-pressure steam line leading from the steam generator to the turbo generator building. The steam temperature will be 1,050°F. The line will be outdoors, where the average ambient temperature is 62°F.

The utility will have a net continuous output capacity of 565 megawatts. Total electrical generation per year is expected to be 2,772 million kWh (plant factor of 56 percent), and the line is expected to be in service 8,500 hours per year. Total expected capital investment for the steam plant (boiler, piping, condenser, etc.) is $55 million. The first year fuel cost will be $18 per ton of coal, which has a heating value of 12,500 Btu/lb. The plant will have a depreciation period of 30 years; however, the insulation is expected to have a useful service life of only 15 years in the outdoor environment. The boiler efficiency is 92 percent, and it requires 10,100 Btu's of fuel to produce each kWh of electricity.

The return on investment requirement for the company is 15 percent. The bonds issued by the company to finance the plant pay a 9 percent dividend, with flotation and administrative costs adding another 1 percent to this cost over the 30-year life. Fuel cost is expected to increase at an average annual rate of 7 percent.

The insulation to be used on the pipe is calcium silicate with an aluminum jacket. The average price estimate per unit by insulation contractors for this insulation on a 16-inch pipe at the time of construction are as follows:

Thickness, inches	$/lin-ft
2" single layer	$11.86
3" single layer	15.39
4" single layer	17.29
5" double layer	27.36
6" double layer	31.79
7" double layer	36.50

Complexity factors, from Table 4-1, are not applied to these numbers because of the absence of many fittings in the pipe.

The worksheets that follow display the input data and resulting economic thickness for the above problem, which is 6 inches. As shown, the calculation was first made using the single layer incremental cost (m_{c1}, = $2.72/in-lin-ft). The thickness using m_{c1} was in the double layer range, so the procedure was repeated using the double layer slope (m_{c2} = $4.57/in-lin-ft). This incremental cost produced the economic thickness of 6 inches, which is in the double layer range and is, therefore, the correct solution.

Cost of Heat Worksheet

1. Multiplier for average annual heat cost, A, using Figure 3-1

   a. Enter Insulation Project Life, years n_1 = __15__ years

   b. Enter Annual Fuel Price Increase i_1 = __0.07__

   c. Find Multiplier A   A = __1.67__

2. First Year Cost of Heat, C_h, using Figure 3-2 (a, b, c or d)

   a. Enter Heating Value of Fuel   H = __12,500__ Btu/

   b. Enter Efficiency of Conversion, fuel to heat   E = __0.92__

   c. Enter First Year Price of Fuel   P(o,g,c,e) = $__18__ / ton

   d. Find First Year Cost of Heat C_h = $__.78__ /$10^6$ Btu

3. Average Annual Heat Cost, using Figure 3-3

   a. Find Average Annual Cost of Heat for purchased steam and electric heat plants (no operating or maintenance costs) AC_h = $_____/$10^6$ Btu

   b. Find Average Annual Heat Cost for coal, oil, and gas plants (10 percent operation and maintenance costs added) $(1.1)AC_h$ = $__1.40__/$10^6$ Btu

4. Compound Interest Factor, $(1+i_2)^{n_2}$, using Figure 3-4

   a. Enter Life of Facility, years n_2 = __30__ years

   b. Enter Annual Cost of Money to finance plant i_2 = __0.10__

   c. Find Compound Interest Factor $(1+i_2)^{n_2}$ = __17.0__

5. Annual Amortization Multiplier for Capital Investment, B, using Figure 3-5

Cost of Heat Worksheet, (Continued)

a. Find B, using i_2 and $(1+i_2)^{n_2}$ from 4, above

B = __.105__

6. Annual Capital Cost of Heat, C_k, using Figure 3-6

   a. Enter Expected Average Annual Heat Production, Q

   $\left(10,100 \frac{Btu}{kWh} \times 2772 \frac{10^6 kWh}{yr} \times 0.92\right) \times 10^{-6}$

   Q = __25.76__ millions of 10^6 Btu

   b. Enter Capital Investment of Heat Plant

   P_f = $__55__ million

   c. Find Annual Capital Cost of Heat

   C_k = $__.23__ 10^6 Btu

7. Find Project Cost of Heat, M,
   M = (1.1)AC_h + C_k

   (1.1)AC_h from step 3 = $__1.40__ /$10^6$Btu

   C_k from step 6 = $__.23__ /$10^6$Btu

   M = $__1.63__ /$10^6$Btu

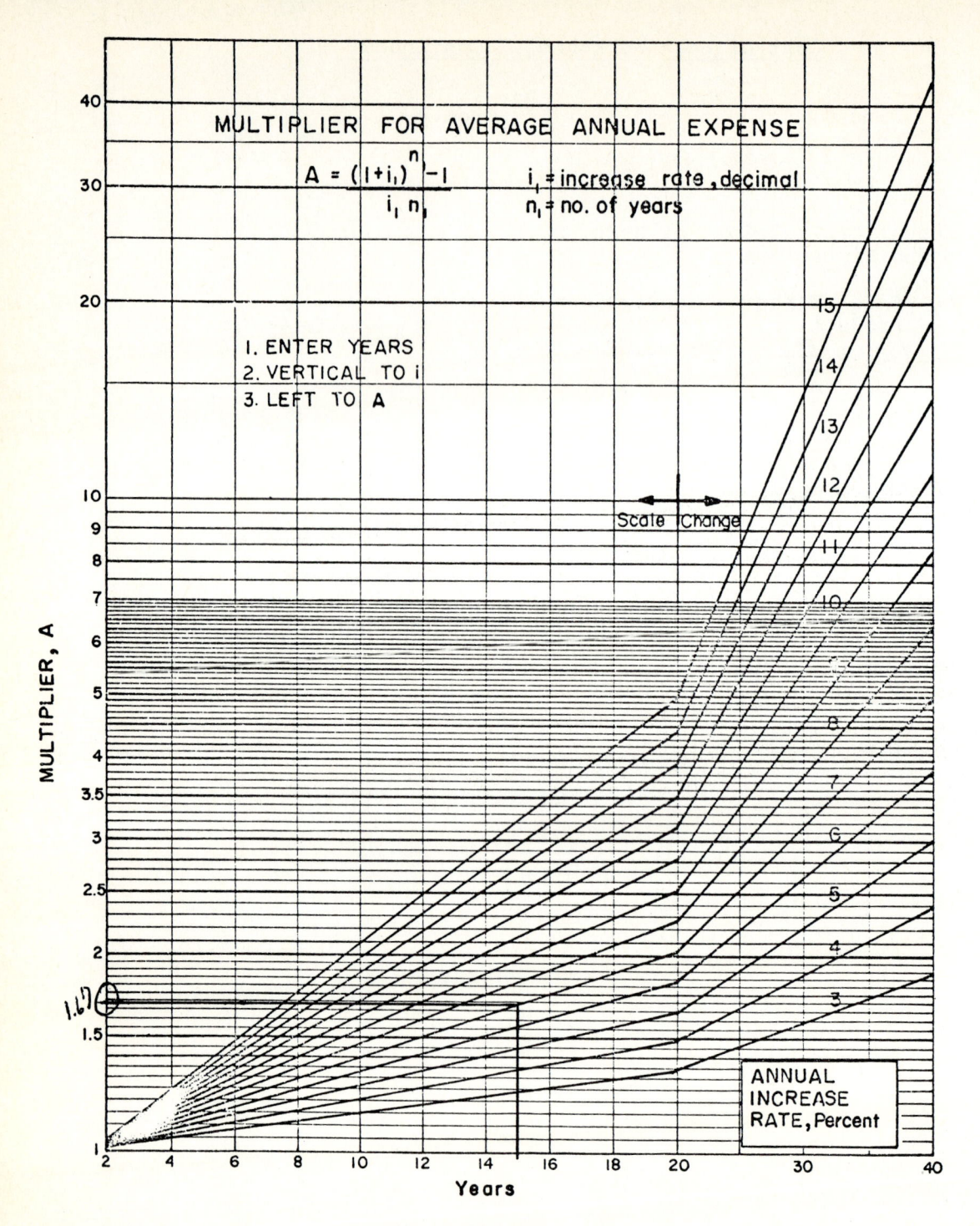

Figure 3-1. Multiplier to apply to present costs for determining the average annual costs when uniform cost increases occur in future years

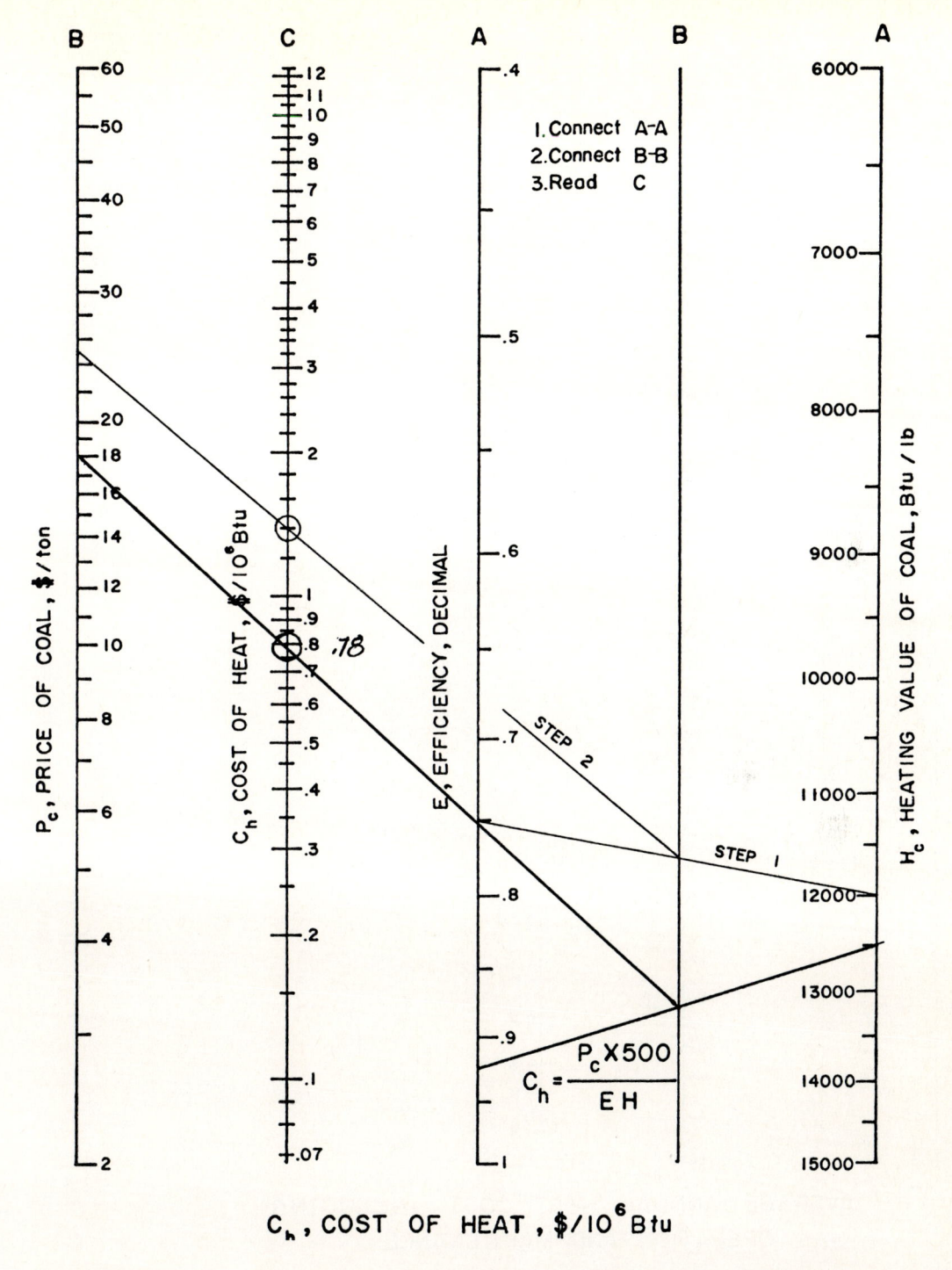

Figure 3-2c. The cost of heat, C_h, for coal as the heat source

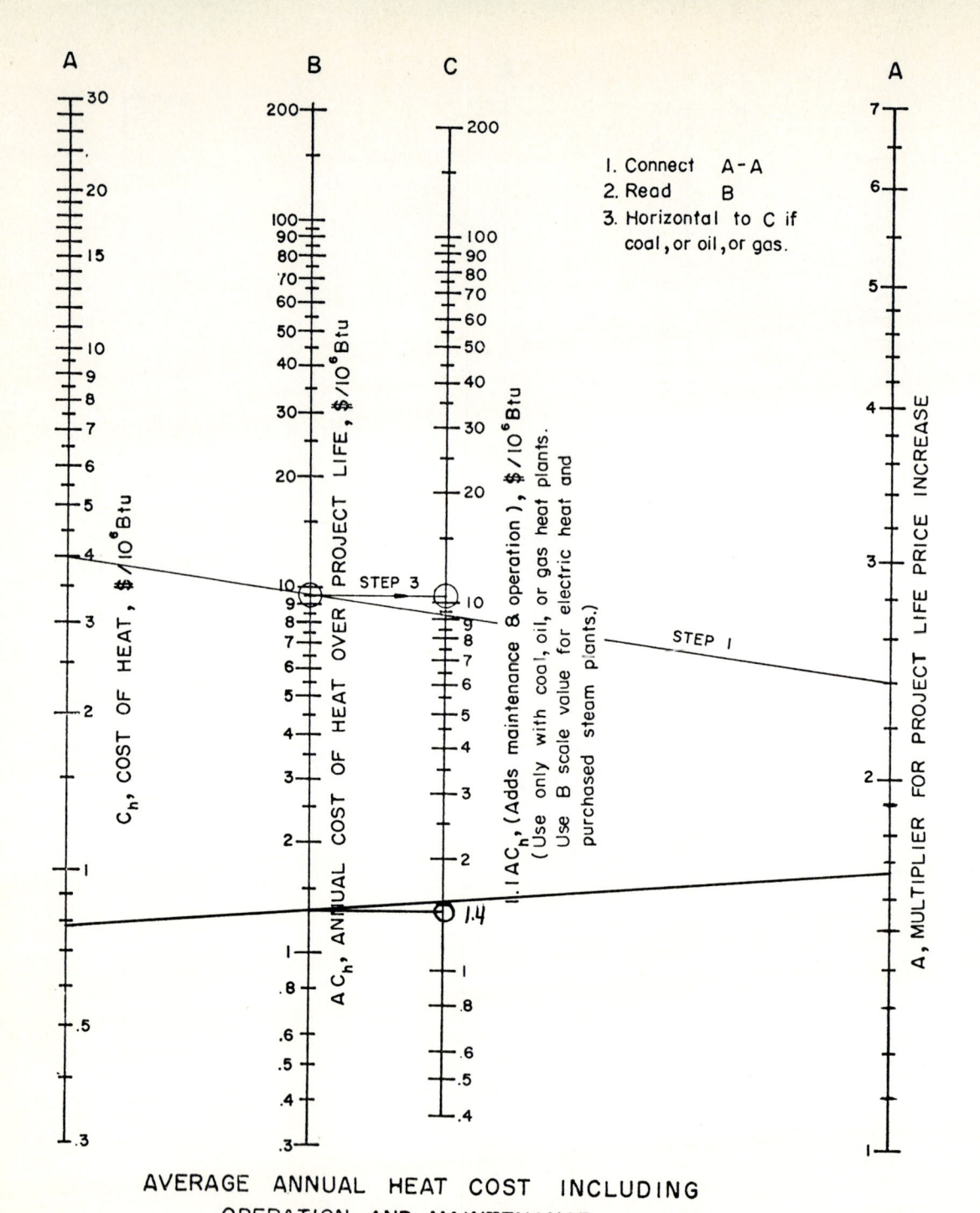

Figure 3-3. The average annual value of heat cost including operation and maintenance costs at the heat producing facility

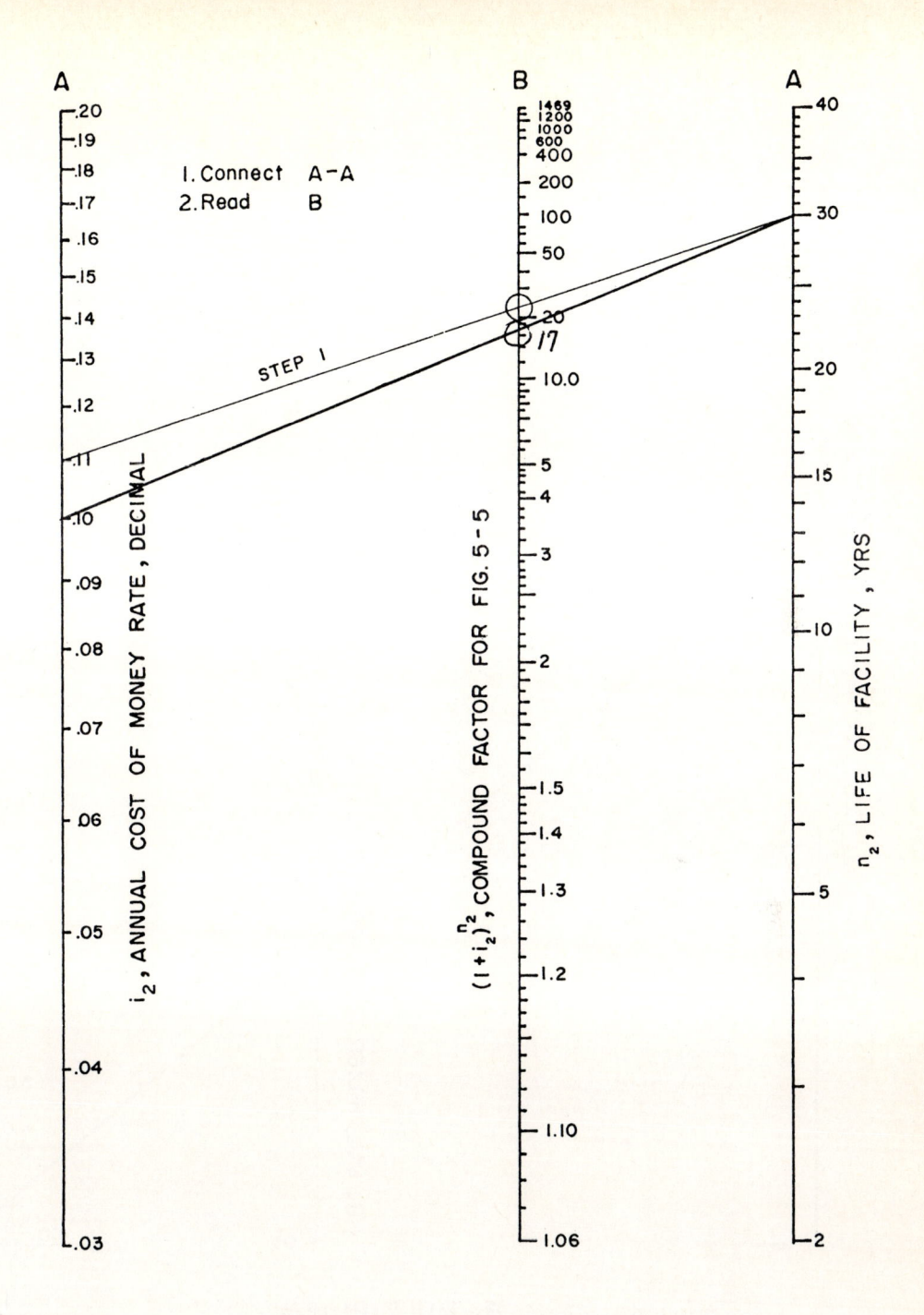

Figure 3-4. Compound interest factor for use with Figure 3-5

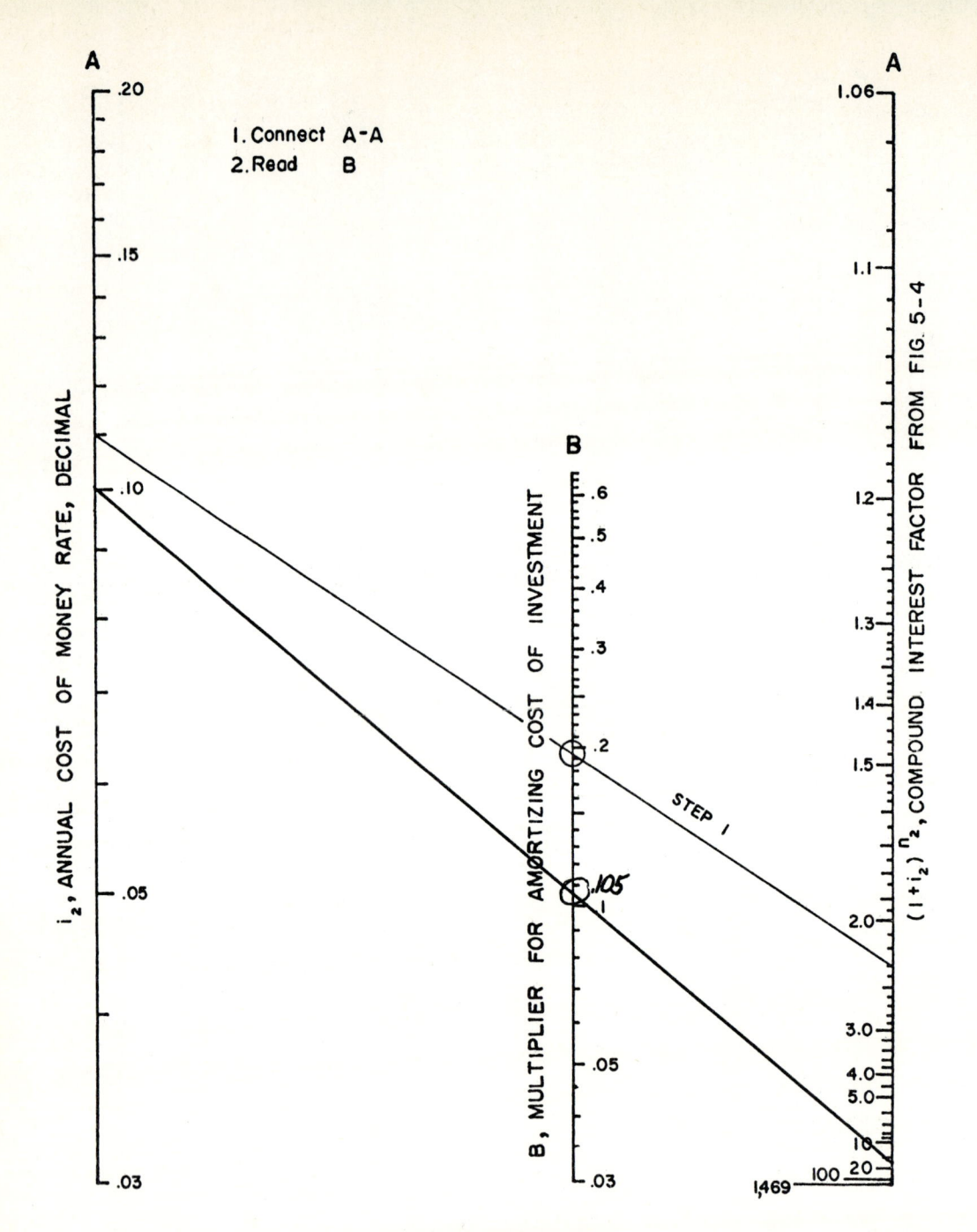

Figure 3-5. The amortization multiplier, B

Note.— The B multiplier $\dfrac{i_2(1+i_2)^{n_2}}{(1+i_2)^{n_2} - 1}$ for amortizing an initial cost over a period of time with equal increments.

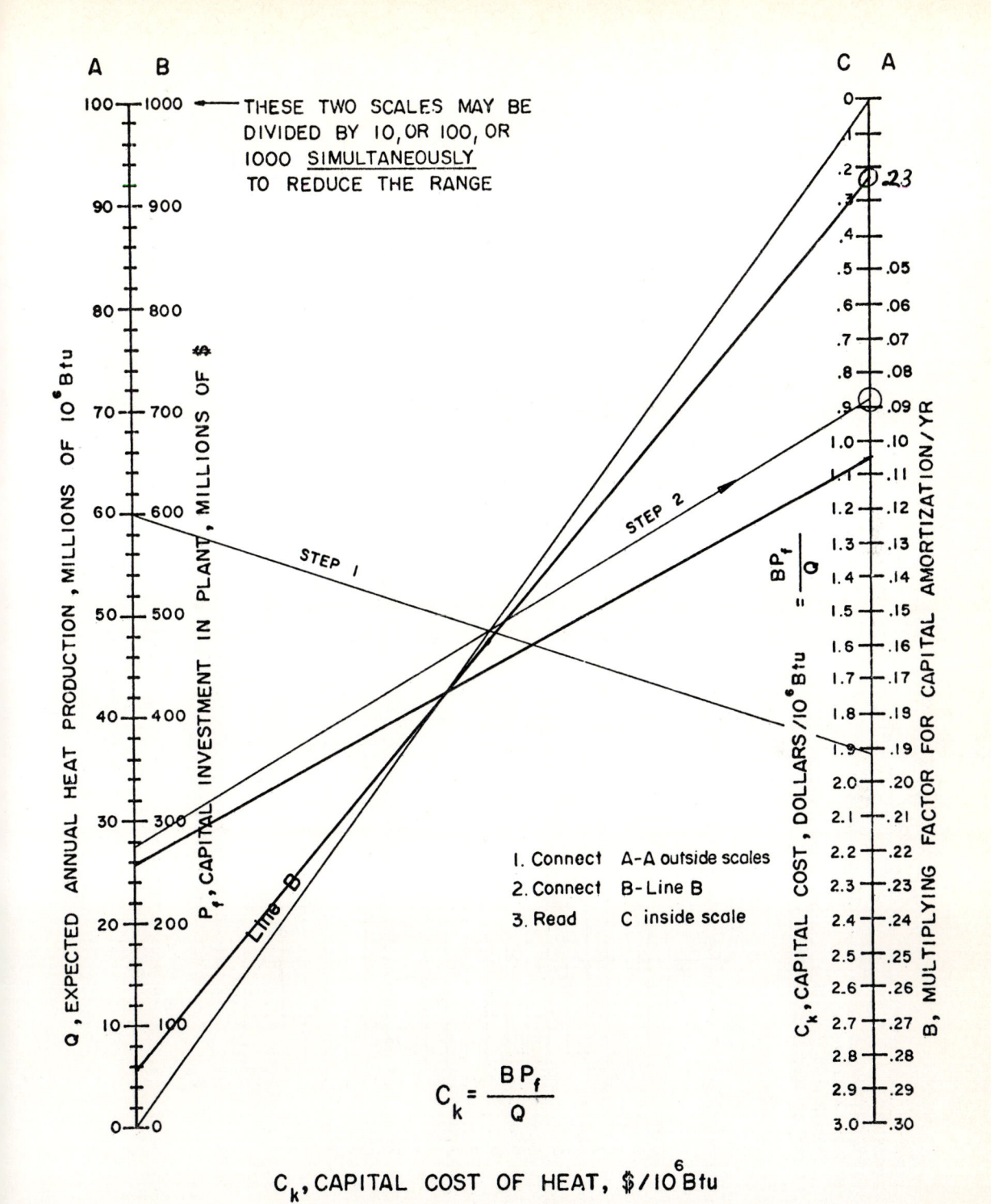

Figure 3-6. The distribution of the heat production capital costs over the energy output on an annual basis

INSULATION COST WORKSHEET

PLANT _Sample Problem #1_ DATE _____

LOCATION _____

APPLICATION _____

SPECIFICATION
 Insulation _____
 Jacket and Finish _____

L - Insulation thickness, inches
P - Installed price
    Piping - $/linear foot
    Vessels - $/ft²
    (obtain from contractor or by using Estimator)
m̄_c - Incremental cost
    Piping - $/lin, ft. per inch
    Vessels - $/ft² per inch

Pipe Size	Single Layer Prices				Double Layer Prices					Triple Layer Prices				
	L_1	P_1	L_2	P_2	m_{c1}	L_1	P_1	L_2	P_2	m_{c2}	L_1	P_1	L_2	P_2	m_{c3}
½															
3/4															
1															
1-1/2															
2															
2-1/2															
3															
4															
5															
6															
8															
10															
12															
14						5	27.36								
16	2	11.86	4	17.29	2.72	5	20.01	7	36.50	4.57					
18															
20															
24															
30															
36															
Flat surfaces and vessels															

NOTES:
1. L_1 = thickness of lower end of layer range; L_2 = thickness of upper end of layer range
   P_1 = installed price for L_1; P_2 = installed price for L_2
2. m_c = PC $(P_2 - P_1)/(L_2 - L_1)$
3. PC = Piping Complexity Factor (Table 4-1)

110

Economic Thickness Determination Worksheet
Hot and Cold* Systems

1. Calculate Mean Insulation Temperature.

   $\dfrac{t_p + t_a}{2}$ $\dfrac{1050 + 62}{2}$ $t_m = $ __556__ °F

2. Enter Insulation Thermal conductivity using Figs. 5-1 or 5-2. $k = $ __.61__ $\dfrac{\text{Btu-in}}{\text{hr-ft}^2\text{-°F}}$

3. Calculate Temperature Difference $(t_p - t_a)$. $\Delta T = $ __988__ °F

4. Enter Annual Hours of Operation. $Y = $ __8500__ hrs

5. Find D_s for flat or D_p for pipe using Fig. 5-3. $D_s = $ _____ $D_p = $ __4.3__

6. Determine B_3 using Fig. 5-4 and 5-5 using i_3 money cost for insulation and n_1 insulation project life. $i_3 = $ __0.15__ $n_1 = $ __15__ yrs $B_3 = $ __.17__

7. Use m_{c1} from Chapter 4. $m_{c1} = $ __2.72__ $m_{c2} = $ __4.57__

8. Find Z_s using Fig. 5-6 for flat or Z_p using Fig. 5-7 for pipe. $Z_s = $ _____ $Z_p = $ __8.4__   __5.1__

9. Calculate kR_s $k \times 0.7$ (R_s value) of 0.7 is typical) $kR_s = $ __.43__

10. Use proper Fig. 5-8 to determine economic thickness. $w_1 = $ __8"__   __6"__

*If a cold system, see Chapter 6 also.

Flat	_____
1/2"x	_____
3/4"x	_____
1" x	_____
1-1/2"x	_____
2" x	_____
2-1/2"x	_____
3" x	_____
4"x	_____
5"x	_____
6"x	_____
8"x	_____
10"x	_____
12"x	_____
14"x	_____
16"x	__6"__
18"x	_____
20"x	_____
24"x	_____
30"x	_____
36"x	_____

11. If the economic thickness found in Step 10 is within the single layer range (corresponding to the single layer slope, m_{c1}, used in Step 7), the thickness is correct. If the thickness is beyond the single layer range, repeat the procedure from Step 7 on, using the double layer slope, m_{c2}.

$$w_2 = \underline{\qquad 6'' \qquad}$$

12. If the economic thickness, using the double layer slope, is in the triple layer range, repeat the procedure from Step 7 on, using the triple layer slope, m_{c3}.

$$w_3 = \underline{\qquad\qquad\qquad}$$

Note.—It may happen that the economic thickness found with a single layer slope falls in the double layer realm and that the subsequent thickness found with a double layer slope falls in the single layer realm; or that the economic thickness found with a double layer slope falls in the triple layer realm and the subsequent thickness found with a triple layer slope falls in the double layer realm. Should either of these conditions occur, the proper economic choice is the thickest single or thickest double layer, respectively.

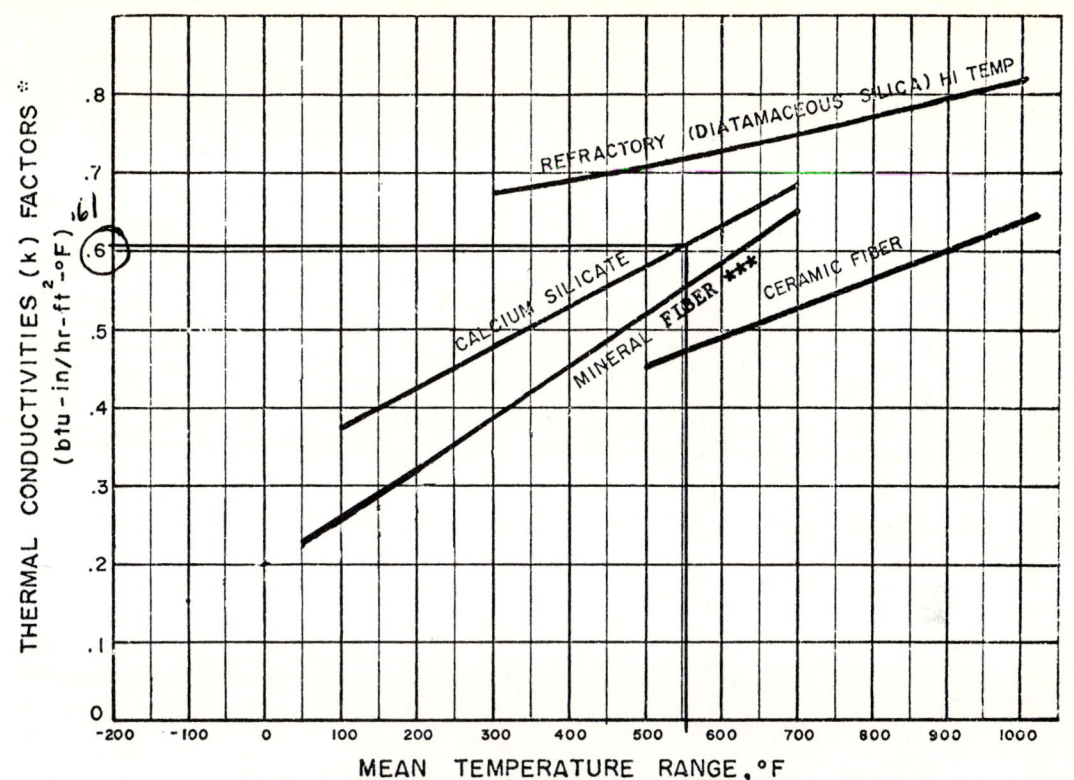

Figure 5-1. High temperature insulation materials

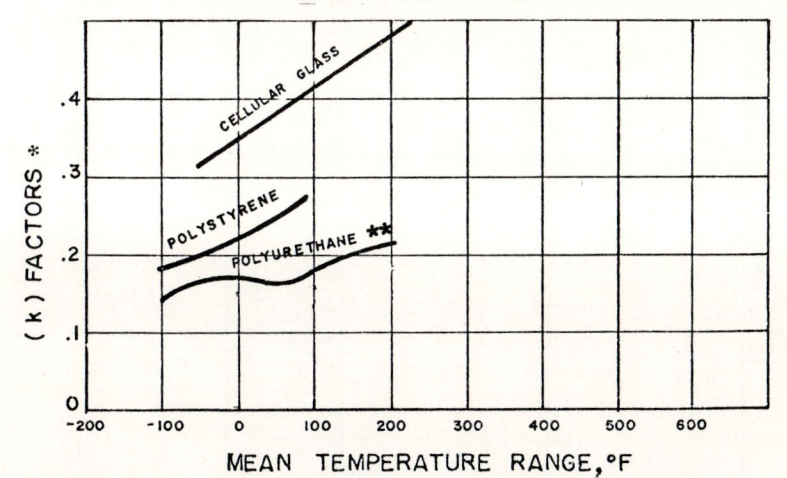

Figure 5-2. Low temperature insulation materials

*The determination of in-service values has yet to be determined for most materials. Should the user have accurate k values for his choice of material those values should be used.

**New material values. Increase by 25 percent for aged value.

***Mineral fiber means rock wool, slag wool, and glass fiber.

113

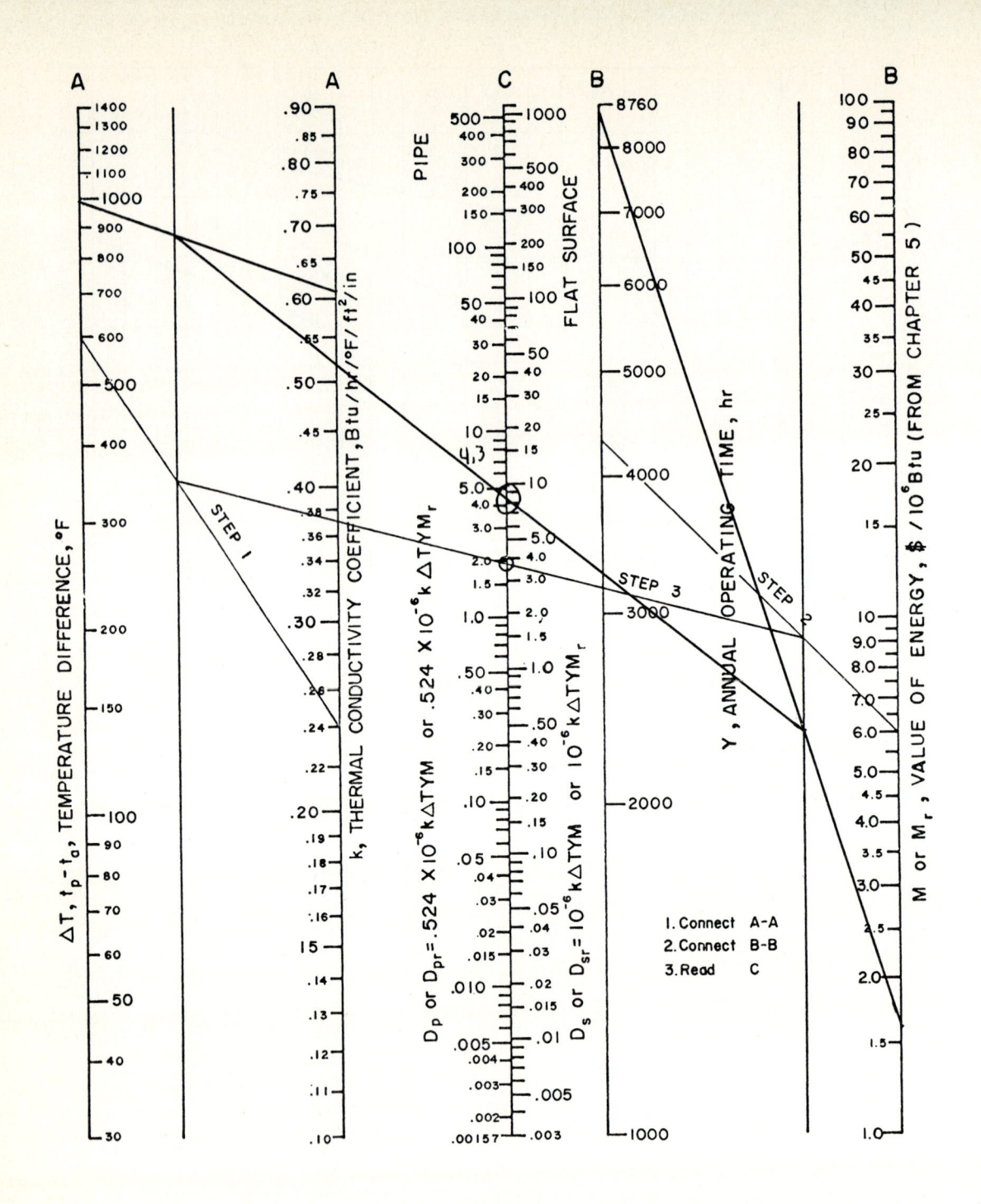

Figure 5-3. Annual cost of heat lost or gained

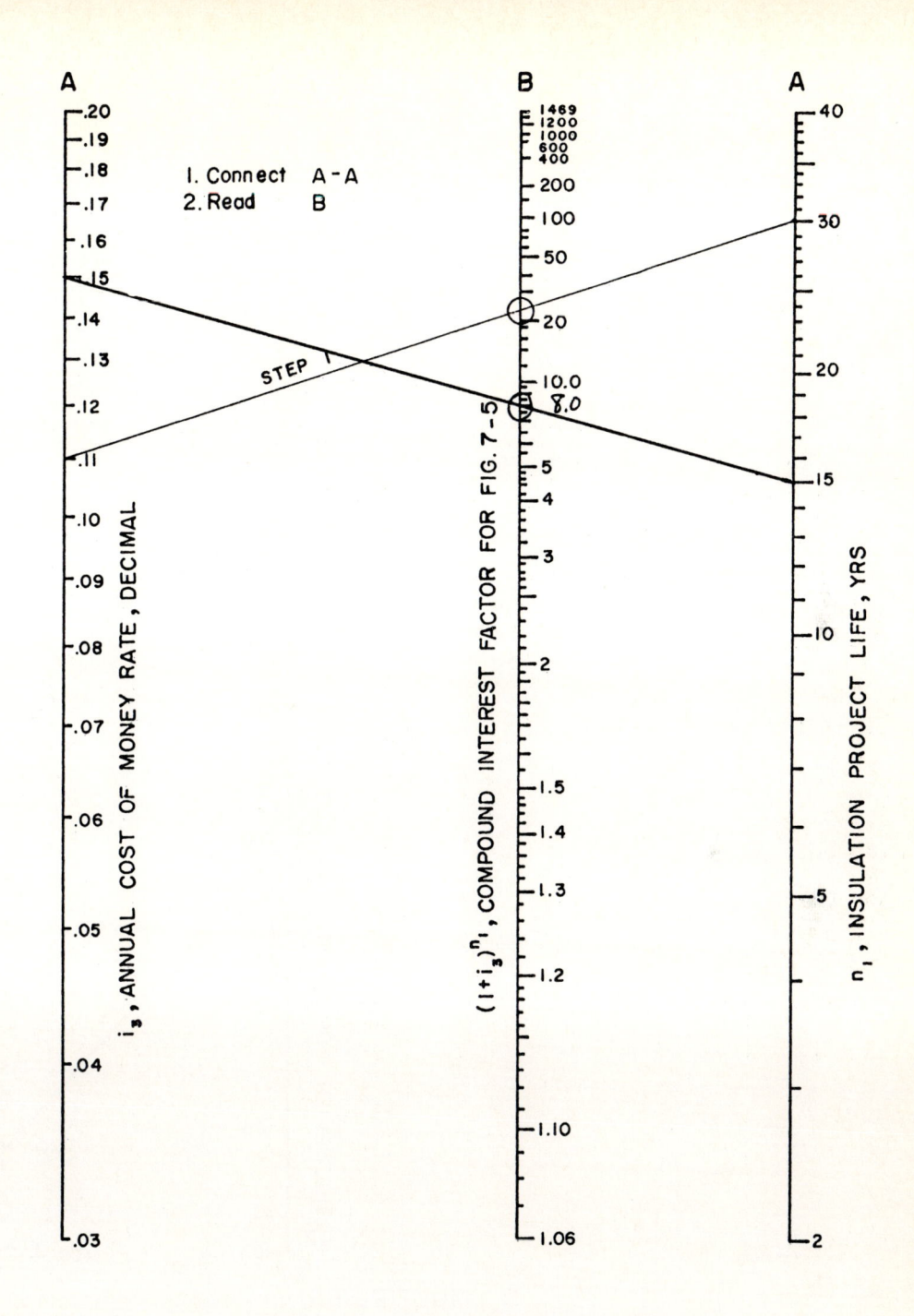

Figure 5-4. Compound interest factor for use with Figure 5-5

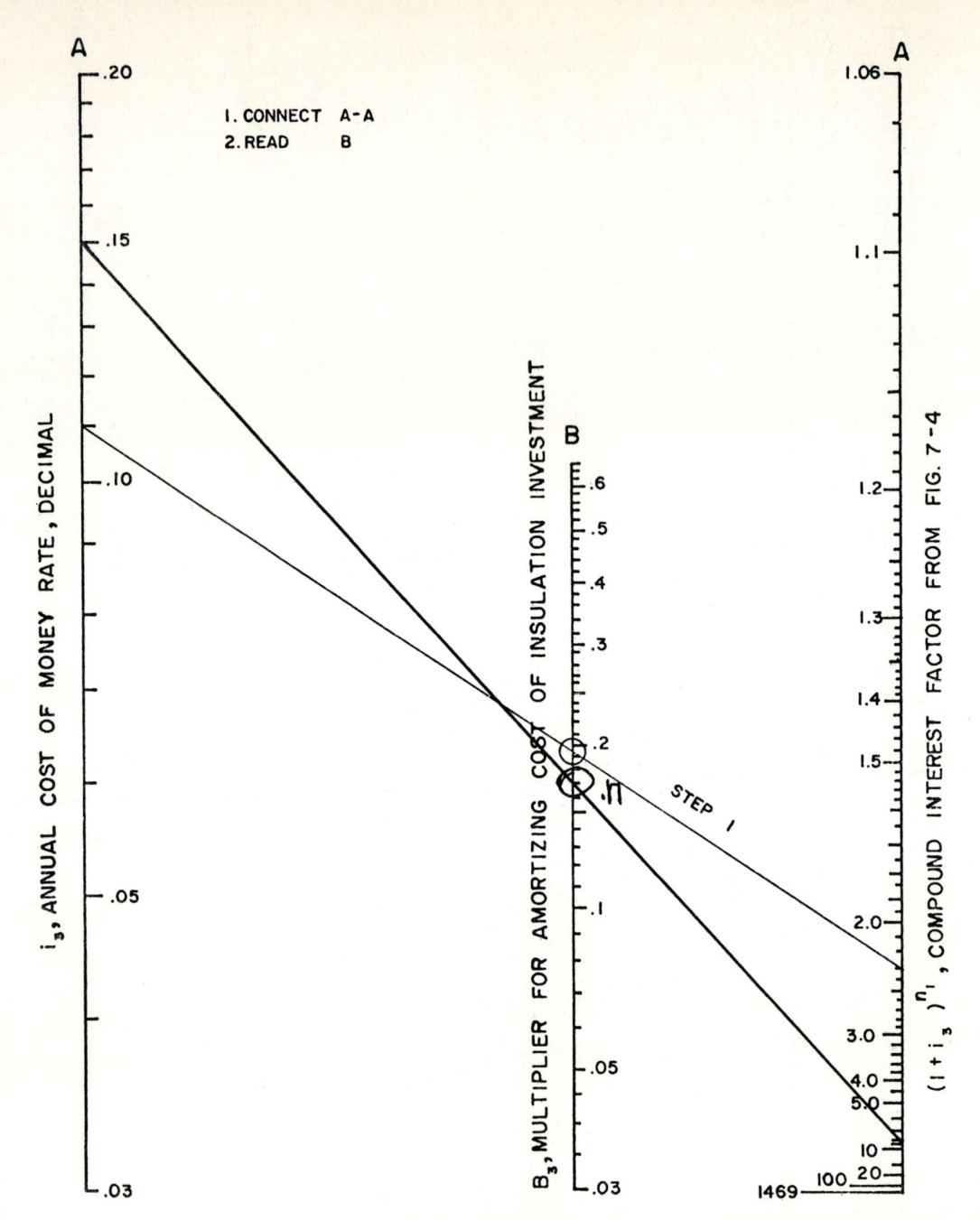

B₃, MULTIPLIER

The B₃ multiplier $\frac{i(1+i)^n}{(1+i)^n - 1}$ for amortizing an initial cost over a period of time with equal increments.

Figure 5-5. <u>The amortizing multiplier, B</u>

Figure 5-7. Z_p, factor for round (pipe) surface

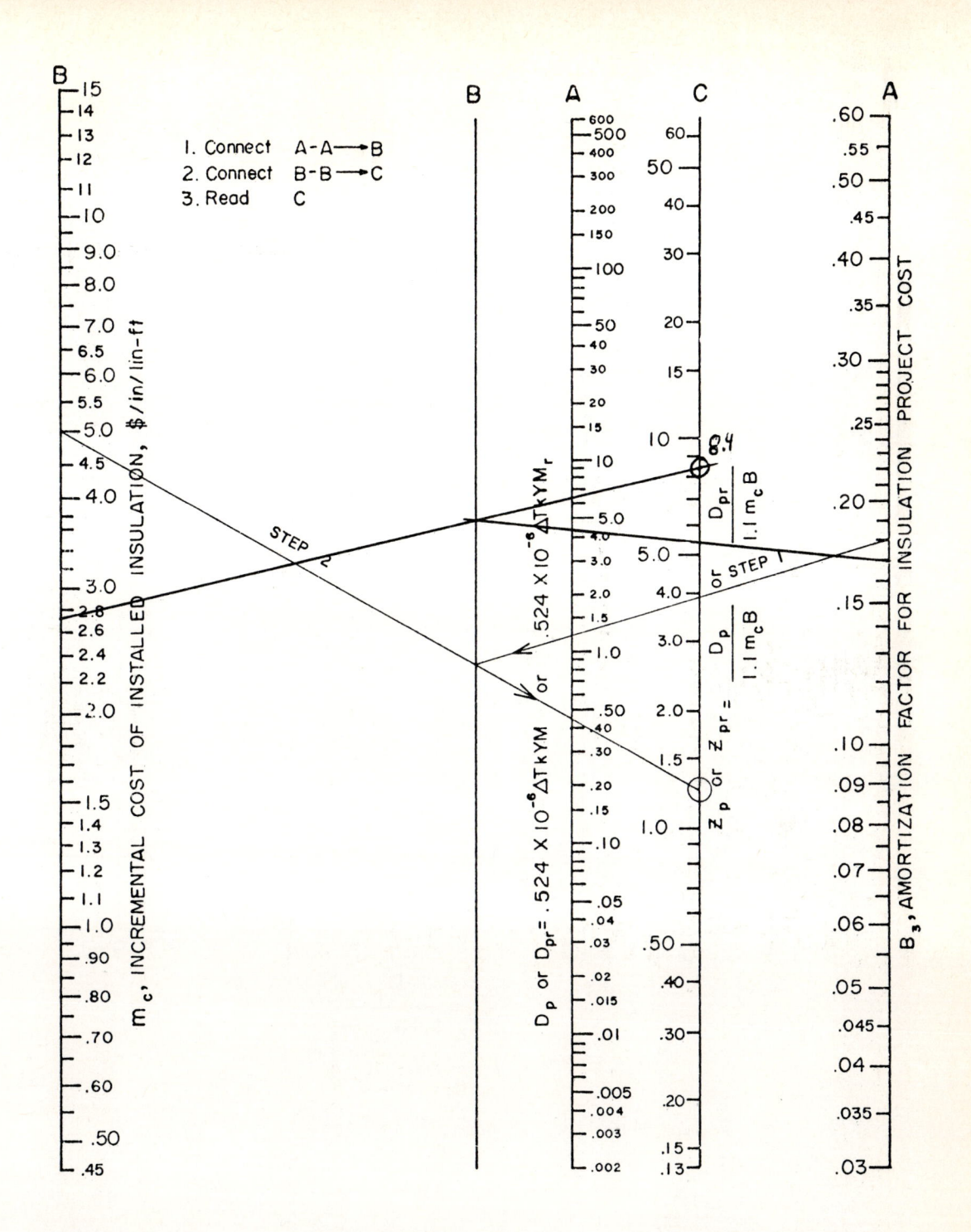

Note.- Z_p includes 10 percent annual maintenance charge for insulation. See Appendix A.

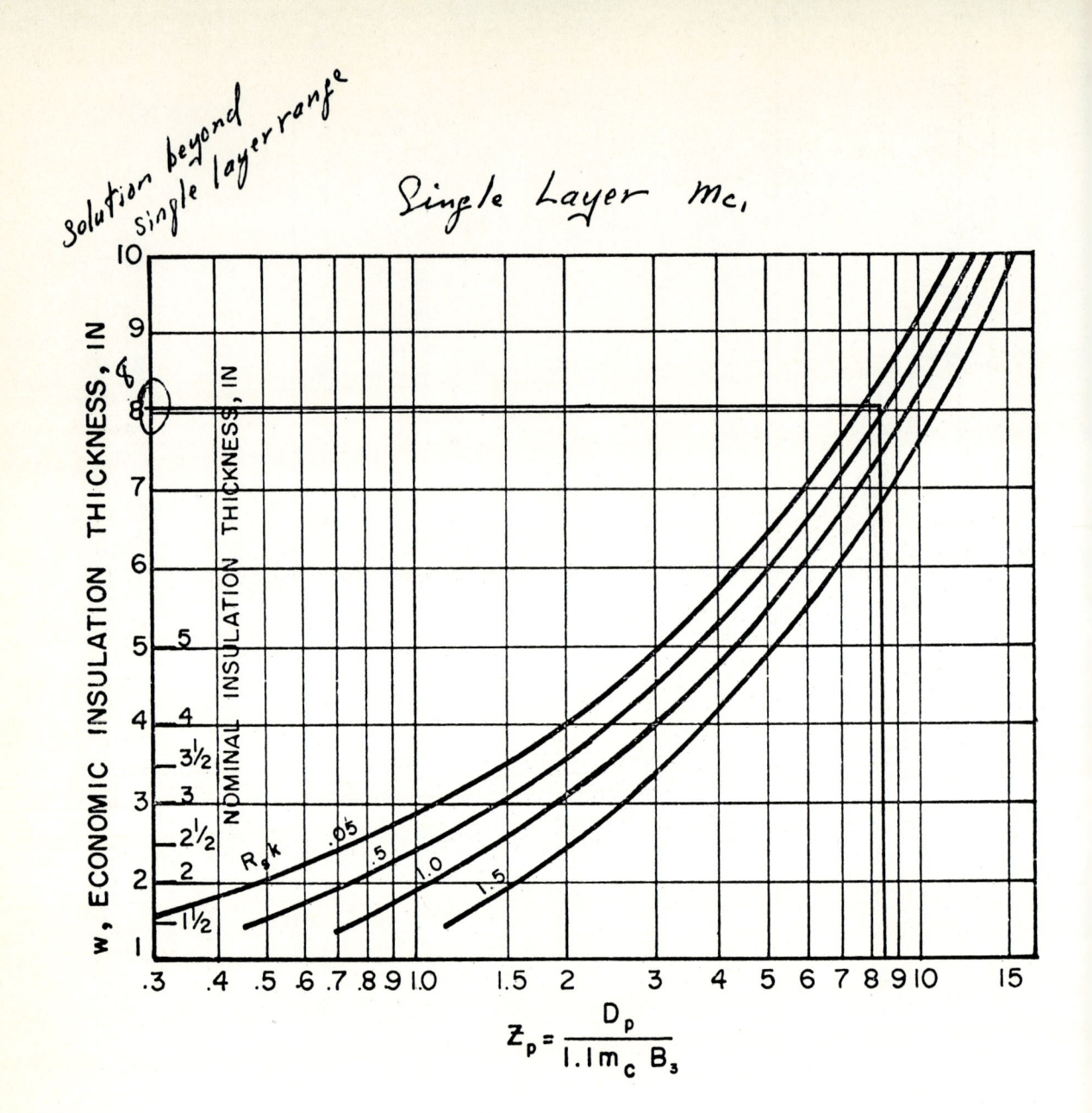

Figure 5-8p. w, Economic thickness, 16-inch pipe

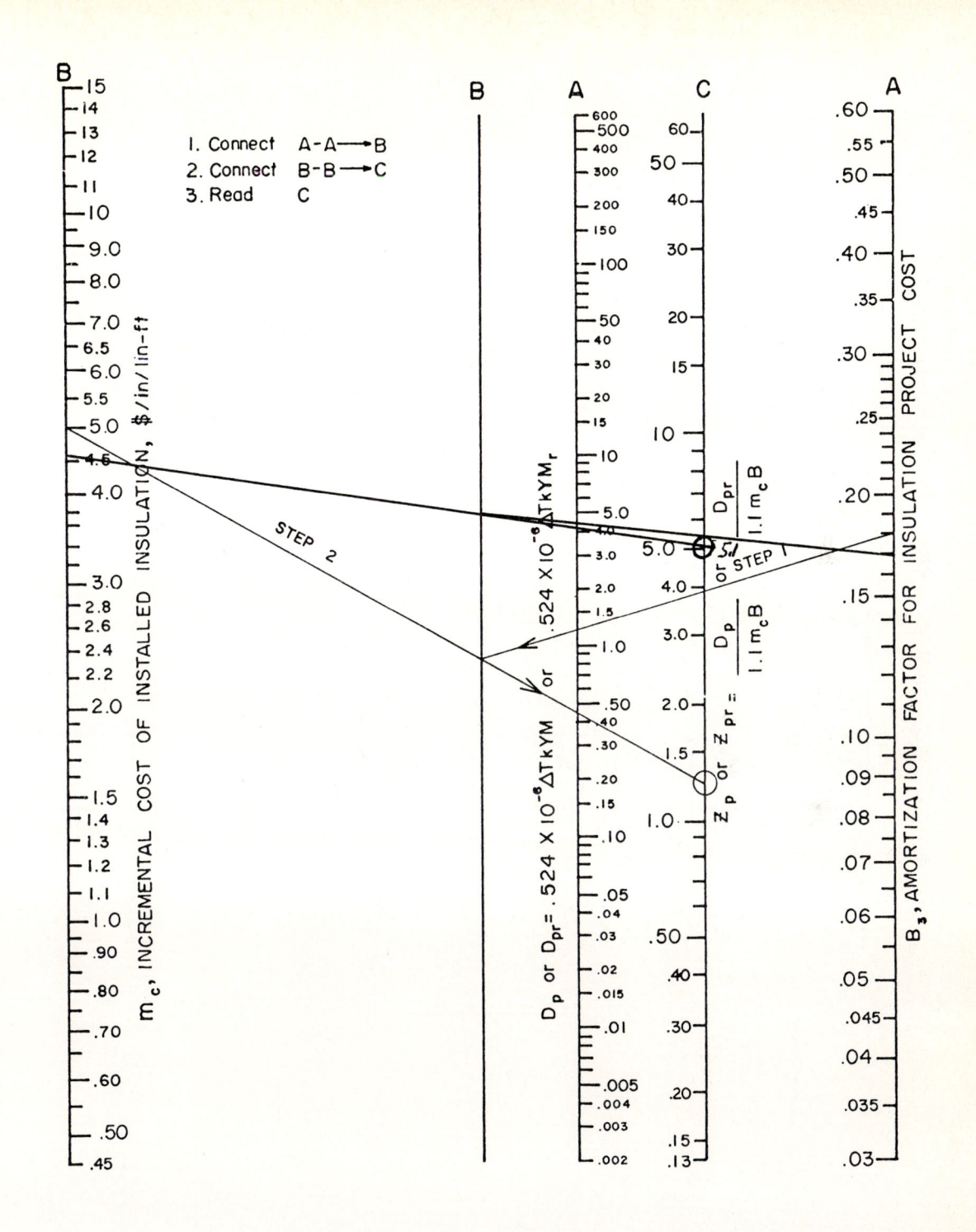

Note.— Z_p includes 10 percent annual maintenance charge for insulation. See Appendix A.

Figure 5-7. Z_p, factor for round (pipe) surface

Double layer Mez

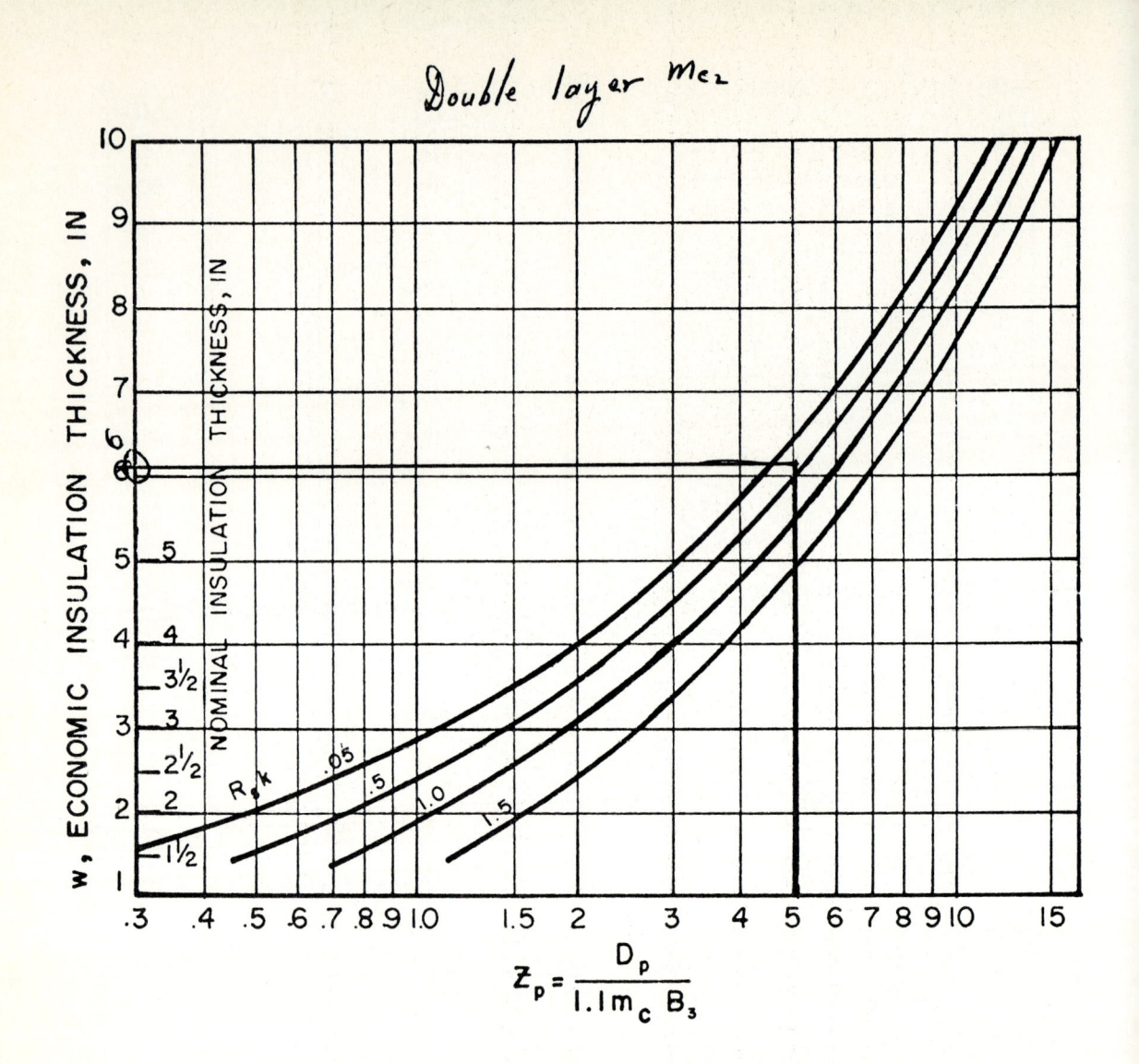

Figure 5-8p. w, Economic thickness, 16-inch pipe

$$Z_p = \frac{D_p}{1.1 m_c B_3}$$

No. 2-Low-Temperature Steam Pipe and Chilled Water Line

A 6-inch NPS line will carry 280°F-steam to an absorption refrigeration unit. The chiller will send out 35°F-water in an 8 inch NPS line to a large computer center for air conditioning. The plant is to be located in Houston, Texas, and is expected to be in operation continuously.

The boiler will be fueled with natural gas, which will cost $1.90 per 1,000 ft^3 at time of plant start up, and increase in price at an expected average annual rate of 8 percent over the 15-year life of the insulation. The natural gas will have a heating value of 1,000 Btu/ft^3, and a conversion efficiency of 80 percent. The capital investment for the heating plant will be $200,000.

The cost of money for the insulation project is 10 percent. The facility will be depreciated over 20 years, with the facility cost of money also being 10 percent. The coefficient of performance for the absorption unit is 0.6, and the cooling load will average 150 tons. The absorption unit will have a water cooled condenser, with water costing $0.0005 per gallon at plant start-up, with an expected annual cost increase of 3 percent. The absorption unit associated chilled water lines, condenser, etc., will cost $100,000.

The insulation on both the steam and cold water lines will be fiber glass. The cold water insulation will have a factory applied vapor barrier, and will first be specified based on economics. This thickness will then be checked for condensation control using design criteria for Houston. Average ambient temperature is 80°F. The average installed costs for insulation as obtained from contractor bids are as follows:

Thickness, inches	6-inch pipe	8-inch pipe
1"	$2.78/lin-ft	3.91
2"	4.03	5.78
3"	5.60	7.74
4"	7.24	9.83

Cost of Heat Worksheet

1. Multiplier for average annual heat cost, A, using Figure 3-1

    a. Enter Insulation Project Life, years n_1 = __15__ years

    b. Enter Annual Fuel Price Increase i_1 = __0.08__

    c. Find Multiplier A   A = __1.81__

2. First Year Cost of Heat, C_h, using Figure 3-2 (a, b, c or d)

    a. Enter Heating Value of Fuel   H = __1000__ Btu/ft^3

    b. Enter Efficiency of Conversion, fuel to heat   E = __0.80__

    c. Enter First Year Price of Fuel   P(o,g,c,e) = $__1.90__/

    d. Find First Year Cost of Heat C_h = $__2.40__/$10^6$ Btu

3. Average Annual Heat Cost, using Figure 3-3

    a. Find Average Annual Cost of Heat for purchased steam and electric heat plants (no operating or maintenance costs) AC_h = $~~4.75~~/$10^6$ Btu

    b. Find Average Annual Heat Cost for coal, oil, and gas plants (10 percent operation and maintenance costs added)   (1.1)AC_h = $__4.75__/$10^6$ Btu

4. Compound Interest Factor, $(1+i_2)^{n_2}$, using Figure 3-4

    a. Enter Life of Facility, years n_2 = __20__ years

    b. Enter Annual Cost of Money to finance plant i_2 = __0.10__

    c. Find Compound Interest Factor $(1+i_2)^{n_2}$ = __6.70__

5. Annual Amortization Multiplier for Capital Investment, B, using Figure 3-5

Cost of Heat Worksheet, (Continued)

    a. Find B, using i_2 and $(1+i_2)^{n_2}$ from 4, above   B = .117

6. Annual Capital Cost of Heat, C_k, using Figure 3-6

    a. Enter Expected Average Annual Heat Production, Q   Q = .0263 millions of 10^6 Btu

$$\frac{150 \text{ ton} \times 12000 \frac{Btu}{ton}}{.6 \text{ COP}} \times 8760 \frac{hr}{yr} \times 10^{-12}$$

    b. Enter Capital Investment of Heat Plant P_f = $ .20 million

    c. Find Annual Capital Cost of Heat C_k = $ .89 10^6 Btu

7. Find Project Cost of Heat, M,
M = (1.1)AC_h + C_k (1.1)AC_h from step 3 = $ 4.75 /10^6Btu

    C_k from step 6 = $ .89 /10^6Btu

    M = $ 5.64 /10^6Btu

INSULATION COST WORKSHEET

PLANT *Sample Problem #2*
LOCATION
APPLICATION
SPECIFICATION
  Insulation
  Jacket and Finish

DATE _____

L – Insulation thickness, inches
P – Installed price
    Piping – $/Linear foot
    Vessels – $/ft²
    (obtain from contractor or by using Estimator)
$\bar{m}_c$ – Incremental cost
    Piping – $/lin. ft. per inch
    Vessels – $/ft² per inch

Pipe Size	Single Layer Prices					Double Layer Prices					Triple Layer Prices				
	L_1	P_1	L_2	P_2	$\bar{m}_{c1}$	L_1	P_1	L_2	P_2	$\bar{m}_{c2}$	L_1	P_1	L_2	P_2	$\bar{m}_{c3}$
½															
3/4															
1															
1-1/2															
2															
2-1/2															
3															
4															
5															
6	1	2.78	4	7.24	1.86										
8	1	3.91	4	9.84	2.27										
10															
12															
14															
16															
18															
20															
24															
30															
36															
Flat surfaces and vessels															

NOTES:
1. L_1 = thickness of lower end of layer range; L_2 = thickness of upper end of layer range
2. P_1 = installed price for L_1; P_2 = installed price for L_2
3. $\bar{m}_c$ = PC $(P_2 - P_1)/(L_2 - L_1)$
   PC = Piping Complexity Factor (Table 4-1)

6" *Complexity factor = 1.25*
8"   " = 1.15

124

Sample Problem #2
Economic Thickness Determination Worksheet
Hot and Cold* Systems
6" steam line

1. Calculate Mean Insulation Temperature.

   $\dfrac{t_p + t_a}{2} \quad \dfrac{250+80}{2}$ $t_m =$ __180__ °F

2. Enter Insulation Thermal conductivity using Figs. 5-1 or 5-2. $k =$ __.29__ $\dfrac{\text{Btu-in}}{\text{hr-ft}^2\text{-°F}}$

3. Calculate Temperature Difference $(t_p - t_a)$. $\Delta T =$ __200__ °F

4. Enter Annual Hours of Operation. $Y =$ __8760__ hrs

5. Find D_s for flat or D_p for pipe using Fig. 5-3. $D_s =$ _____ $D_p =$ __1.5__

6. Determine B_3 using Fig. 5-4 and 5-5 using i_3 money cost for insulation and n_1 insulation project life.
   $i_3 =$ __0.10__
   $n_1 =$ __15__ yrs
   $B_3 =$ __.13__

7. Use m_{cl} from Chapter 4. $m_{cl} =$ __1.86__

8. Find Z_s using Fig. 5-6 for flat or Z_p using Fig. 5-7 for pipe. $Z_s =$ _____ $Z_p =$ __5.6__

9. Calculate kR_s $k \times 0.7$ (R_s value of 0.7 is typical) $kR_s =$ __.203__

10. Use proper Fig. 5-8 to determine economic thickness. $w_1 =$ __4"__

*If a cold system, see Chapter 6 also.

Flat	_____
1/2"x	_____
3/4"x	_____
1"x	_____
1-1/2"x	_____
2"x	_____
2-1/2"x	_____
3"x	_____
4"x	_____
5"x	_____
6"x	__4"__
8"x	_____
10"x	_____
12"x	_____
14"x	_____
16"x	_____
18"x	_____
20"x	_____
24"x	_____
30"x	_____
36"x	_____

11. If the economic thickness found in Step 10 is within the single layer range (corresponding to the single layer slope, m_{c1}, used in Step 7), the thickness is correct. If the thickness is beyond the single layer range, repeat the procedure from Step 7 on, using the double layer slope, m_{c2}.

$w_2 = $ _____

12. If the economic thickness, using the double layer slope, is in the triple layer range, repeat the procedure from Step 7 on, using the triple layer slope, m_{c3}.

$w_3 = $ _____

Note.—It may happen that the economic thickness found with a single layer slope falls in the double layer realm and that the subsequent thickness found with a double layer slope falls in the single layer realm; or that the economic thickness found with a double layer slope falls in the triple layer realm and the subsequent thickness found with a triple layer slope falls in the double layer realm. Should either of these conditions occur, the proper economic choice is the thickest single or thickest double layer, respectively.

Cost of Refrigeration Worksheet

1. Operating Cost

    a. Determine M, using procedure on Cost of Heat Worksheet

    M=$ _5.64_/10^6Btu

    b. Determine Coefficient of Performance using Figure 4-7

    COP=_.6 (given)_

    c. Find Operating Cost of Refrigeration using Figure 4-8

    M/COP=$ _9.40_/10^6Btu

2. Make-up Water Cost

    a. Find Average Annual Cost Multiplier using Figure 4-1 when

    i_1 = expected annual increase in water price

    i_1= _0.03_

    n_1 = insulation project life

    n_1= _15_ years

    Find average Annual Cost Multiplier

    A= _1.24_

    b. Find S using Table 4-1

    S= _6.2_

    c. Select First Year Price of Water

    P_w=$_.0005_/gal

    d. Find Average Make-up Water Cost, using Figure 4-9

    C_w=$ _.30_ /10^6Btu

3. Refrigeration Equipment Capital Cost Assigned to Cost of Refrigeration, C_{kr}, using Figure 4-10

    a. Enter Annual Operating Time, hrs/yr

    Y= _8760_ hrs/yr

    b. Enter Load, tons

    T_r= _150_ tons

    c. Find Annual Refrigeration, 1,000's ton-hr

    YT_r=_1,320,000_ ton-hr

    d. Enter System Capital Cost, 1,000's $

    P_r=$_100_ 1,000's

    e. Use B from Cost of Heat Worksheet Step 5

    B= _.117_

    f. Find C_{kr}, Capital Cost of Refrigeration

    C_{kr}=$ _.74_/10^6Btu

Cost of Refrigeration Worksheet, (Continued)

4. Find M_r

   $M_r = M/COP + C_w + C_{kr}$

   M/COP from Step 1 = $ 9.40 /10⁶Btu
   C_w from Step 2 = $ .30 /10⁶Btu
   C_{kr} from Step 3 = $ .74 /10⁶Btu
   M_r = $ 10.44 /10⁶Btu

Sample problem #2
Economic Thickness Determination Worksheet
Hot and Cold* Systems
8" Chilled Water Line

1. Calculate Mean Insulation Temperature.

   $\dfrac{t_p + t_a}{2}$ $t_m =$ __57.5__ °F

2. Enter Insulation Thermal conductivity using Figs. 5-1 or 5-2.

   $k =$ __.23__ $\dfrac{\text{Btu-in}}{\text{hr-ft}^2\text{-}°F}$

3. Calculate Temperature Difference $(t_p - t_a)$. $\Delta T =$ __45__ °F

4. Enter Annual Hours of Operation. $Y =$ __8760__ hrs

5. Find D_s for flat or D_p for pipe using Fig. 5-3.

   $D_s =$ _____
   $D_p =$ __.49__

6. Determine B_3 using Fig. 5-4 and 5-5 using i_3 money cost for insulation and n_1 insulation project life.

   $i_3 =$ __0.10__
   $n_1 =$ __20__ yrs
   $B_3 =$ __.117__

7. Use m_{c1} from Chapter 4. $m_{c1} =$ __2.27__ /in-lin-ft

8. Find Z_s using Fig. 5-6 for flat or Z_p using Fig. 5-7 for pipe.

   $Z_s =$ _____
   $Z_p =$ __1.65__

9. Calculate kR_s ($k \times 0.7$ (R_s value) of 0.7 is typical) $kR_s =$ __.161__

10. Use proper Fig. 5-8 to determine economic thickness. $w_1 =$ __2½"__

*If a cold system, see Chapter 6 also.

Flat	_____
1/2"x	_____
3/4"x	_____
1"x	_____
1-1/2"x	_____
2"x	_____
2-1/2"x	_____
3"x	_____
4"x	_____
5"x	_____
6"x	_____
8"x	2½"
10"x	_____
12"x	_____
14"x	_____
16"x	_____
18"x	_____
20"x	_____
24"x	_____
30"x	_____
36"x	_____

Condensation Control Worksheet

1. Ambient Design Air Conditions*

   a. Using Table 6-1 choose design dry bulb temperature. t_a = __94__, °F

   b. Using Table 6-1 choose design wet bulb temperature. t_w = __80__, °F

   c. Using Table 6-2 find design relative humidity. rh = 0.__54__

   d. Using Fig. 6-1 find design dewpoint temperature. t_d = __75__, °F

   e. Determine design minimum surface temperature. $t_d + 1 = t_s$ t_s = __76__, °F

2. Allowable Heat Gain, Q_a

   a. Find surface emissivity (Table 6-3). ε = __.63__

   b. Find radiation heat gain using Fig. 6-2. Q_r = __13__ Btu/hr-ft²

   c. Find convective heat gain using Fig. 6-3. Q_{cv} = __11__ Btu/hr-ft²

   d. Determine total allowable heat gain ($Q_a = Q_r + Q_{cv}$). Q_a = __24__ Btu/hr-ft²

3. Insulation Thickness

   a. Enter process temperature. t_p = __35__, °F

   b. Find insulation k value using Fig. 5-2 (Chapter 5). k = __.23__ $\frac{\text{Btu-in}}{\text{hr-ft}^2\text{-°F}}$

   c. Find equivalent insulation thickness using Fig. 6-4. L = __½__, in

   d. Find actual pipe and flat surface thickness using Fig. 6-5. w_c = __½__, in

*Should the user have local data, that data should be used.

No. 3-Retrofit Condensate Line/Payback Analysis

The plant engineer for a paper and pulp factory is considering insulating 5-inch, 170°F-steam condensate return line that was not insulated when the plant was built 5 years ago. The ambient temperature averages 70°F. Although he is concerned about saving energy, the plant engineer doubts that any insulation applied would pay for itself in the required 3 years on a discounted cash flow basis, as is a company policy for energy conservation modifications.

The steam generator average output is 200,000 pounds of steam per hour. The steam temperature is 370°F, and pressure is 160 psig. The boiler is in operation 8,400 hours per year. The boiler plant cost $3.5 million when built 5 years ago, with the cost of money for the facility being 7 percent. The boiler efficiency is 88 percent, and the original depreciation period of the boiler was 20 years. The fuel used is No. 2 oil, which presently costs $0.30 per gallon. Fuel price is expected to increase at an annual rate of 6 percent.

The cost of money for the insulation investment is 8 percent. The type of insulation to be used is fiber glass. Estimated installed costs are as follows:

Thickness, inches	$/lin ft
1"	$3.91
1½"	4.48
2"	5.09

Sample Problem #3
Cost of Heat Worksheet

1. Multiplier for average annual heat cost, A, using Figure 3-1

    a. Enter Insulation Project Life, years $\quad n_1 =$ __3__ years

    b. Enter Annual Fuel Price Increase $\quad i_1 =$ __0.06__

    c. Find Multiplier A $\quad A =$ __1.06__

2. First Year Cost of Heat, C_h, using Figure 3-2 (a, b, c or d)

    a. Enter Heating Value of Fuel $\quad H =$ __142,000__ Btu/

    b. Enter Efficiency of Conversion, fuel to heat $\quad E =$ __0.88__

    c. Enter First Year Price of Fuel $\quad P(o,g,c,e) = \$$ __.30__ /

    d. Find First Year Cost of Heat $\quad C_h = \$$ __2.40__ $/10^6$ Btu

3. Average Annual Heat Cost, using Figure 3-3

    a. Find Average Annual Cost of Heat for purchased steam and electric heat plants (no operating or maintenance costs) $\quad AC_h = \$$ _____ $/10^6$ Btu

    b. Find Average Annual Heat Cost for coal, oil, and gas plants (10 percent operation and maintenance costs added) $\quad (1.1) AC_h = \$$ __2.80__ $/10^6$ Btu

4. Compound Interest Factor, $(1+i_2)^{n_2}$, using Figure 3-4

    a. Enter Life of Facility, years $\quad n_2 =$ __20__ years

    b. Enter Annual Cost of Money to finance plant $\quad i_2 =$ __0.07__

    c. Find Compound Interest Factor $\quad (1+i_2)^{n_2} =$ __3.87__

5. Annual Amortization Multiplier for Capital Investment, B, using Figure 3-5

Cost of Heat Worksheet, (Continued)

    a. Find B, using i_2 and $(1+i_2)^{n_2}$ from 4, above   B = .094

6. Annual Capital Cost of Heat, C_k, using Figure 3-6

    a. Enter Expected Average Annual Heat Production, Q *   Q = 2.01 millions of 10^6 Btu

    b. Enter Capital Investment of Heat Plant P_f = $3.5 million

    c. Find Annual Capital Cost of Heat C_k = $ .16 10^6 Btu

7. Find Project Cost of Heat, M,
$M = (1.1)AC_h + C_k$

     $(1.1)AC_h$ from step 3 = $ 2.80 /10^6 Btu

     C_k from step 6 = $ .16 /10^6 Btu

     M = $ 2.96 /10^6 Btu

* $\dfrac{Btu}{yr} = 200{,}000 \dfrac{lb\ steam}{hr} \times 1196 \dfrac{Btu^{**}}{lb\ steam} \times 8400 \dfrac{hr}{yr}$

$\dfrac{Btu}{yr} \times 10^{-12} = Q$ millions of $10^6 \dfrac{Btu}{yr}$

** Steam @ 370°F and 160 psig has a heat content (enthalpy) of 1196 $\dfrac{Btu}{lb}$, as found in Table 3-3.

PLANT _Sample Problem 3_
LOCATION _____
APPLICATION _____
SPECIFICATION _____
Insulation _____
Jacket and Finish _____

INSULATION COST WORKSHEET
DATE _____

L – Insulation thickness, inches
P – Installed price
    Piping – $/Linear foot
    Vessels – $/ft²
    (obtain from contractor or by using Estimator)
$\bar{m}_c$ – Incremental cost
    Piping – $/lin. ft. per inch
    Vessels – $/ft² per inch

Pipe Size	Single Layer Prices					Double Layer Prices					Triple Layer Prices				
	L_1	P_1	L_2	P_2	m_{c1}	L_1	P_1	L_2	P_2	m_{c2}	L_1	P_1	L_2	P_2	m_{c3}
½															
3/4															
1															
1-1/2															
2															
2-1/2															
3															
4															
5	1	3.91	2	5.09	1.36										
6															
8															
10															
12															
14															
16															
18															
20															
24															
30															
36															
Flat surfaces and vessels															

NOTES:
1. L_1 = thickness of lower end of layer range; L_2 = thickness of upper end of layer range
    P_1 = installed price for L_1; P_2 = installed price for L_2
2. $\bar{m}_c$ = PC $(P_2 - P_1)/(L_2 - L_1)$
3. PC = Piping Complexity Factor (Table 4-1)

5" complexity factor
= 1.15

Sample Problem 3
Economic Thickness Determination Worksheet
Hot and Cold* Systems

1. Calculate Mean Insulation Temperature.

   $\dfrac{t_p + t_a}{2} \quad \dfrac{170+70}{2} \quad t_m = \underline{\quad 120 \quad}$ °F

2. Enter Insulation Thermal conductivity using Figs. 5-1 or 5-2. $k = \underline{\quad .28 \quad} \dfrac{\text{Btu-in}}{\text{hr-ft}^2\text{-°F}}$

3. Calculate Temperature Difference $(t_p - t_a)$. $\Delta T = \underline{\quad 100 \quad}$ °F

4. Enter Annual Hours of Operation. $Y = \underline{\quad 8400 \quad}$ hrs

5. Find D_s for flat or D_p for pipe using Fig. 5-3.
   $D_s = \underline{\quad\quad}$
   $D_p = \underline{\quad .36 \quad}$

6. Determine B_3 using Fig. 5-4 and 5-5 using i_3 money cost for insulation and n_1 insulation project life.
   $i_3 = \underline{\quad 0.08 \quad}$
   $n_1 = \underline{\quad 3 \quad}$ yrs
   $B_3 = \underline{\quad .385 \quad}$

7. Use m_{cl} from Chapter 4. $m_{cl} = \underline{\quad 1.36 \quad}$

8. Find Z_s using Fig. 5-6 for flat or Z_p using Fig. 5-7 for pipe.
   $Z_s = \underline{\quad\quad}$
   $Z_p = \underline{\quad .63 \quad}$

9. Calculate kR_s $k \times 0.7$ (R_s value) of 0.7 is typical) $kR_s = \underline{\quad .196 \quad}$

10. Use proper Fig. 5-8 to determine economic thickness. $w_1 = \underline{\quad 1'' \quad}$

*If a cold system, see Chapter 6 also.

Flat	_____
1/2"x	_____
3/4"x	_____
1" x	_____
1-1/2"x	_____
2" x	_____
2-1/2"x	_____
3" x	_____
4"x	_____
5"x	1"
6"x	_____
8"x	_____
10"x	_____
12"x	_____
14"x	_____
16"x	_____
18"x	_____
20"x	_____
24"x	_____
30"x	_____
36"x	_____

BIBLIOGRAPHY

1. American Gas Association, Department of Statistics, Gas Facts: A Statistical Record of the Gas Utility Industry in 1973.

2. ASHRAE, Guide and Data Book, Equipment, 1969.

3. ASHRAE, Guide and Data Book, Systems, 1970.

4. ASHRAE, Guide and Data Book, Applications, 1971.

5. ASHRAE, Handbook of Fundamentals, 1972.

6. ASTM, Book of Standards, Part 18, Thermal Insulation, 1975.

7. ASTM, Thermal Insulation in the Petrochemical Industry, 1974, STP 581.

8. Carmichael, C., Kents's Mechanical Engineers' Handbook Design and Production, 12th edition, (John Wiley and Sons, New York), 1950.

9. Federal Construction Council, Insulation of Mechanical Systems, Federal Construction Guide Specification, Section 1516, Feb. 1972.

10. Federal Power Commission, Steam Electric Plant Construction Cost and Annual Production Expenses, 25th annual, 1972 (U.S. Government Printing Office).

11. Foster Associates, Inc., Energy Prices 1960-1973 (Ballinger Publishing Co., Cambridge, Mass.), 1974.

12. Gibbons, A. M., Insulation Technology and the Contractor: Selected Advances in Insulation Technology (Small Business Administration), 1970.

13. Glaser, P. E., et al., Thermal Insulation Systems - A Survey, A. D. Little for NASA (National Technical Information Service), 1967.

14. Gyftopoulos, E. P., et al., Potential Fuel Effectiveness in Industry (Ballinger Publishing Co., Cambridge, Mass.), 1974.

15. Herendeen, R. A., The Energy Cost of Goods and Services, ORNL NSF EP-58 (National Technical Information Service), Oct. 1973.

16. Keenan, J. and J. Kaye, Gas Tables (John Wiley and Sons, New York), 1945.

17. Malloy, J. F., Thermal Insulation (Van Nostrand Rheinhold, New York), 1969.

18. Martin, R. B., "Guide to better insulation," Chemical Engineering, (82), May 12, 1975.

19. Modern Plastics Encyclopedia, Volume 45, No. 14A, (McGraw-Hill, New York), 1968.

20. National Academy of Sciences, National Research Council, Thermal Insulation Thickness Charts, Publication 1084, (Federal Construction Council Technical Report No. 45), 1963.

21. National Bureau of Standards, Thermal Resistance of Air Spaces and Fibrous Insulation Bounded by Reflective Surfaces (U.S. Department of Commerce), 1957.

22. National Climatic Center, Monthly Normals of Temperature, Precipitation, and Heating and Cooling Degree Days 1941-70, Climatography of the U.S. No. 81 (by State)(U.S. Department of Commerce).

23. National Insulation Manufacturers Association, Economic Thickness of Insulation for Flat Surfaces and Pipes in the Low Temperature Range, 1963.

24. National Insulation Manufacturers Association, How to Determine Economic Thickness Insulation, July, 1961.

25. North American Manufacturing Co., North American Combustion Handbook, 1st Edition (Cleveland, Ohio), 1965.

26. Oak Ridge National Laboratory: unpublished draft of insulation material report prepared for ERDA.

27. O'Keefe, William, "Thermal insulation," Power, (118), August 1974.

28. Ottaviano, V. B., National Mechanical Estimator 1973, (Ottaviano Technical Services, Melville, L.I.).

29. Perry, J. H., Chemical Engineer's Handbook, 4th edition, (McGraw-Hill, New York), 1950.

30. Salisbury, J. Kenneth, Kent's Mechanical Engineer's Handbook, 12th edition, (John Wiley & Sons, Inc.), 1950.

31. Selby, S. H., Standard Mathematical Tables, 21st edition, (Chemical Rubber Co.), 1973.

32. Shreve, R. N., Chemical Process Industries, 3rd edition, (McGraw-Hill, New York), 1967.

33. Shuman, E. C., "Thermal insulation systems," Modern Materials, (7), 1970.

34. Snow, Robert, Building Construction Cost Data, 32d edition, (Means Co., Inc., Duxbury, Mass.), 1974.

35. Stanford Research Institute, Patterns of Energy Consumption in the United States, (National Technical Information Service, U.S. Department of Commerce, PB 212-776), Feb. 1972.

36. Underwood, W. F., "Advances in reflection insulation allow tighter control of costs," Power, (119), June 1975.

37. U.S. Atomic Energy Commission, Power Plant Capital Costs, Current Trends, and Sensitivity to Economic Parameters, (WASH-1345, U.S. Government Printing Office), Oct. 1974.

38. U.S. Department of Commerce, Bureau of Census, Statistical Abstracts of the U.S., 1974, 95th edition.

39. U.S. Department of Commerce and Federal Energy Administration, Retrofitting Existing Housing for Energy Conservation: An Economic Analysis, (Building Science Series 64), December 1974.

40. Williams, Franklin E., "Thermal insulation in building construction," Construction Review, November 1973.

41. Wilson, A. C., Industrial Thermal Insulation, (New York, McGraw-Hill), 1959.

42. York Research Corporation, ECON-I, How to Determine Economic Thickness of Thermal Insulation, (Thermal Insulation Association, Mt. Kisco, New York), 1973.

APPENDIX A: DERIVATION OF ECONOMIC THICKNESS EQUATIONS

The cost of heat lost through the insulation covering a pipe is derived as follows.

Heat Flow

From any standard text or handbook the flow of heat is directly related to the temperature increment across an increment of insulation thickness, the coefficient of thermal conductivity for the insulation material and the area of the material. The relation is

$$q_p = -KA \frac{dT}{dR} \qquad \text{A-1}$$

where: q_p = the flow, $\frac{Btu}{hr}$

K = the coefficient, $\frac{Btu\text{-}ft}{hr\text{-}ft^2\text{-}°F}$

A = the area, ft^2

dT = the incremental temperature, $°F$

dR = the incremental radius (thickness), ft

Since this is a pipe the insulation material forms a shell around the pipe. The shell length will be given as L, ft and the shell area will then be

$A = 2\pi RL$ at the radius R, ft for which the incremental thickness dR is considered.

Thus

$$q_p = -2\pi K\, RL\, \frac{dT}{dR} \qquad \text{A-2}$$

Rearranging and applying integration across

$$q_p \int_{R_1}^{R_2} \frac{dR}{R} = -2\pi KL \int_{T_1}^{T_2} dT \qquad \text{A-3}$$

the insulation shell from the inside radius R_1 to the outside radius R_2 and between the inside temperature T_1 and the outside temperature T_2 we have

$$q_p \ln \frac{R_2}{R_1} = -2\pi KL (T_2-T_1) \qquad \text{A-4}$$

Let $(T_2-T_1) = \Delta T$ and rearrange

$$q_p = \frac{-2\pi KL \Delta T}{\ln \frac{R_2}{R_1}} \quad , \quad \frac{Btu}{hr} \qquad \text{A-5}$$

Now we divide by A_2 to get heat flow/unit outside area where $A_2 = 2\pi R_2 L$

$$\frac{q_p}{A_2} = \frac{-2\pi KL \Delta T}{2\pi R_2 L \ln \frac{R_2}{R_1}}$$

simplifying

$$\frac{q_p}{A_2} = \frac{-K \Delta T}{R_2 \ln \frac{R_2}{R_1}} \quad , \quad \frac{Btu}{hr\text{-}ft^2} \qquad \text{A-6}$$

At this point it is convenient to shift into an inches scale:

$$K, ft \rightarrow k, in$$
$$R_2, ft \rightarrow r_2, in$$
$$R_1, ft \rightarrow r_1, in$$

so that

$$\frac{q_p}{A_2} = \frac{-k \Delta T}{r_2 \ln \frac{r_2}{r_1}} = \frac{Btu}{hr\text{-}ft^2} \qquad \text{A-7}$$

If we rearrange A-7, thusly,

$$\frac{q_p}{A_2} = \frac{-\Delta T}{\frac{r_2}{k} \ln \frac{r_2}{r_1}}$$

we convert the equation 7 into a thermal resistance form which accounts only for the heat flow resistance from the insulation.

There is also a surface resistance through which the heat must flow. To account for that resistance a term R_s is added to the equation 7 as

$$\frac{q_p}{A_2} = \frac{-\Delta T}{\frac{r_2}{r_1} \ln \frac{r_2}{r_1} + R_s} \qquad \text{A-8}$$

rearranging we get

$$\frac{q_p}{A_2} = \frac{-k \ \Delta T}{r_2 \ln \frac{r_2}{r_1} + kR_s} \ , \ \frac{Btu}{hr\text{-}ft^2} \qquad \text{A-9}$$

To get $\frac{Btu}{hr\text{-}ft(length)}$, $\frac{q}{A_2} \left\{ \frac{Btu}{hr\text{-}ft^2} \right\} \times \frac{2\pi r_2 \ L}{12 \ L} \left\{ \frac{ft^2}{ft} \right\} =$

$$\frac{-\pi r_2 \ k \ \Delta T}{6(r_2 \ln \frac{r_2}{r_1} + kR_s)} \ , \ \frac{Btu}{hr\text{-}ft}$$

so that the heat flow per foot of pipe is

$$U_p = \frac{-.524 \ k \ \Delta T \ r_2}{r_2 \ln \frac{r_2}{r_1} + kR_s} \ , \ \frac{Btu}{hr\text{-}ft} \qquad \text{A-10}$$

This is, then, the heat lost from a foot of pipe with an insulating jacket over the pipe. (The negative sign merely indicates which way the heat is flowing.)

The assumptions that have been made are as follows:

1. k is the insulation thermal conductivity at the mean temperature across the insulation. That is, k is taken at $\frac{T_2 + T_1}{2}$ as it was measured by the ASTM-C177 test method.

2. ΔT is the temperature of the fluid inside the pipe, T_2, less the temperature of the ambient air, T_1. This ignores the temperature drop across the pipe.

Economic Value

The annual economic value of the heat lost from the pipe is the quantity lost per hour for the annual hours used and dollars per unit of heat. This is

$$m_p = U_p Y M \times 10^{-6}, \quad \frac{\$}{\text{yr-ft}} \qquad \text{A-11}$$

U_p is given by eqn 10, $\frac{\text{Btu}}{\text{hr-ft}}$

Y is the annual operating hours, $\frac{\text{hr}}{\text{yr}}$

M is the average cost of supplying the lost Btu, $\frac{\$}{10^6 \text{ Btu}}$ during the economic period being examined.

See A-34.

expanding

$$m_p = \frac{-.524 \, \Delta T r_2 \, Y M \times 10^{-6}}{r_2 \ln \frac{r_2}{r_1} + k R_s}, \quad \frac{\$}{\text{yr-ft}} \qquad \text{A-12}$$

Now let

$$D = -.524 \times 10^{-6} \, k \, \Delta T Y M, \quad \frac{\$}{\text{yr-ft}}$$

Then

$$m_p = \frac{D r_2}{r_2 \ln \frac{r_2}{r_1} + k R_s}, \quad \frac{\$}{\text{yr-ft}} \qquad \text{A-13}$$

This represents the annual cost of lost heat in the given system. This cost will be added to the annual cost of the insulation system so that the least annual cost may be determined with the insulation thickness as the single variable.

Insulation Cost

The annual cost of the insulation system is characterized by

$$C_i = 1.1\, m_c B\,(r_2 - r_1) + 1.1\, Bd \qquad \text{A-14}$$

See equation C-1.

The total annual cost of heat lost and insulation applied is the sum of the above two costs or

$$Y_p = m_p + C_i \qquad \text{A-15}$$

expanding

$$Y_p = \frac{Dr_2}{r_2 \ln \frac{r_2}{r_1} + kR_s} + 1.1 m_c B(r_2 - r_1) + 1.1 Bd$$

This equation is illustrated by Figure A-1

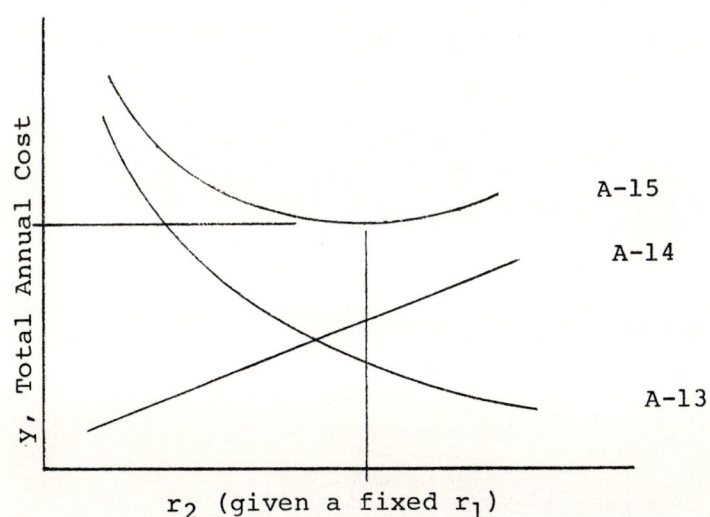

r$_2$ (given a fixed r$_1$)
Total Annual Cost vs. Insulation Thickness

Figure A-1. Basic economic thickness curves

The minimum point of equation 15 can be determined by taking the differential $\frac{dy_p}{dr_2}$ and setting it equal to zero, which will define the most economic r_2, or thickness of insulation.

$$y_p = \frac{Dr_2}{(r_2 \ln \frac{r_2}{r_1} + kR_s)} + 1.1 m_c B(r_2-r_1) + 1.1 Bd \qquad \text{A-16}$$

Take derivative

$$dy_p = \frac{(r_2\ln\frac{r_2}{r_1} + kR_s)D\, dr_2 - Dr_2(r_2 \frac{r_1}{r_2}\frac{dr_2}{r_1} + \ln\frac{r_2}{r_1} dr_2)}{(r_2 \ln \frac{r_2}{r_1} + kR_s)^2} +$$

$$1.1\, m_c B\, dr_2$$

Consolidate

$$\frac{dy_p}{dr_2} = \frac{D\{(r_2\ln\frac{r_2}{r_1} + kR_s) - r_2(1 + \ln\frac{r_2}{r_1})\}}{(r_2\ln\frac{r_2}{r_1} + kR_s)^2} +$$

$$1.1\, m_c B \qquad \text{A-17}$$

Simplify

$$\frac{dy_p}{dr_2} = \frac{D(kR_s - r_2)}{(r_2\ln\frac{r_2}{r_1} + kR_s)^2} + 1.1\, m_c B \qquad \text{A-18}$$

Set equal to 0 and rearrange to identify the minimum point of the curve

$$0 = (kR_s - r_2) + 1.1\, m_c \frac{B}{D}(r_2\ln\frac{r_2}{r_1} + kR_s)^2 \qquad \text{A-19}$$

Let $\frac{1}{Z_p} = \frac{1.1\, m_c B}{D}$, then $(r_2 - kR_s) = \frac{1}{Z_p}(r_2\ln\frac{r_2}{r_1} + kR_s)^2$

and $\quad \dfrac{(r_2\ln\frac{r_2}{r_1} + kR_s)^2}{r_2 - kR_s} = Z_p \qquad \text{A-20}$

144

Equation A-20 is now plotted for each of a number of pipe sizes (r_1) and is the solution to the insulation thickness for least annual cost. Thus, we determine Z_p, enter the curve for the pipe size in question, and read the most economic thickness for whatever kR_s value fits the chosen insulation system. See Figure 5-8 (b through u). There is no compromise of solution with this method. The evaluation of kR_s and Z may, however, contain judgments which the manual user will make and which, of course, will affect the solution.

Before we analyze the makeup of those parameters, we will examine the flat plate solution.

Flat Plate

The fundamental heat transfer equation is the same as before:

$$q_s = -KA \frac{dT}{dW}, \quad \frac{Btu}{hr} \qquad \text{A-21}$$

where

dW is the incremental thickness of the insulation, ft.

Rearrange

$$q_s \int_{W_1}^{W_2} dW = -KA \int_{T_1}^{T_2} dT \qquad \text{A-22}$$

Integrate

$$q_s (W_2 - W_1) = -KA(T_2 - T_1) \qquad \text{A-23}$$

Rearrange

$$\frac{q_s}{A} = \frac{-K \, \Delta T}{W_2 - W_1}, \quad \frac{Btu}{hr\text{-}ft^2}$$

$$= \frac{-\Delta T}{\frac{W_2 - W_1}{K}}, \quad \frac{Btu}{hr\text{-}ft^2} \qquad \text{A-24}$$

Here we change W, ft to w, in and K, $\frac{Btu\text{-}ft}{°F\text{-}hr\text{-}ft^2}$ to k, $\frac{Btu\text{-}in}{°F\text{-}hr\text{-}ft^2}$ as well as letting $w_2 - w_1 = w$, in

Again the surface resistivity needs to be accounted for, so R_s is added

$$\frac{q_s}{A} = \frac{-\Delta T}{\frac{w}{k} + R_s} \quad , \quad \frac{Btu}{hr\text{-}ft^2} \qquad \text{A-25}$$

$$\frac{q_s}{A} = U_s = \frac{-\Delta T}{\frac{w}{k} + R_s} \quad , \quad \frac{Btu}{hr\text{-}ft^2} \qquad \text{A-26}$$

w = thickness of insulation, in

$k = \frac{Btu\text{-}in}{hr\text{-}ft^2\text{-}°F}$

rearrange

$$U_s = \frac{-k\,\Delta T}{w + kR_s} \quad , \quad \frac{Btu}{hr\text{-}ft^2} \qquad \text{A-27}$$

(Assumptions 1 and 2 following A-10 apply to A-27.)

The annual value of heat lost across the surface is then similar to the pipe derivation.

$$m_s = U_s\, YM \times 10^{-6}, \quad \frac{\$}{yr\text{-}ft} \qquad \text{A-28}$$

expanding

$$m_s = \frac{-k\,\Delta TYM \times 10^{-6}}{w + kR_s} \quad , \quad \frac{\$}{yr\text{-}ft^2} \qquad \text{A-29}$$

let $D_s = -k\,\Delta TYM \times 10^{-6}$

then

$$m_s = \frac{D_s}{w + kR_s} \quad , \quad \frac{\$}{yr\text{-}ft^2} \qquad \text{A-30}$$

This is the annual heat loss cost for a square foot of the flat surface with insulation thickness being the independent variable.

We again add the annual cost of installed insulation, which is characterized by

$$C_i = 1.1 m_c B\,w + 1.1\,Bd \qquad \text{A-31}$$

(see C-1.)

to the cost of heat lost

$$Y_s = m_s + C_i$$

or

$$y_s = \frac{D_s}{w + kR_s} + 1.1\, m_c Bw + 1.1 Bd \qquad \text{A-32}$$

Differentiating and setting the differential equal to zero we have,

$$dy_s = \frac{-D_s\, dw}{(w + kR_s)^2} + 1.1\, m_c B dw$$

$$\frac{dy_s}{dw} = \frac{-D_s}{(w + kR_s)^2} + 1.1\, m_c B$$

Let $Z_s = \dfrac{D_s}{1.1 m_c B}$

Rearrange

$$(w + kR_s)^2 = Z_s \qquad \text{A-33}$$

Figure 5-8(a) illustrates A-33. Thus, when Z_s has been determined and kR_s has been chosen, the minimum cost thickness is chosen from the curve.

Equation A-20 Terms

k

The insulation thermal conductivity coefficient, k, is chosen from the insulation type and the mean temperature, t_m, of the application. k values in sales literature are generally lower than conservative design would permit; for this reason, the data given in this manual reflect the ASHRAE "design" values whenever possible.

An error in k of ± 50 percent will result in a nominal +22, -28 percent error in economic thickness. k, therefore, should be evaluated with reasonable care in this determination. (See Fig. B-2.)

t_m, the mean temperature is defined as the cold side temperature of the insulation plus the hot side temperature divided by two. This manual uses $t_m = \dfrac{t_p + t_a}{2}$, which is the process temperature, plus the ambient temperature divided by two. The temperature drops across the pipe or vessel being insulated and the air surface at the insulation offer little or no change to the t_m for determining k and are, therefore, not considered for simplicity in the analysis.

R_s

The thermal resistance at the insulation surface is a function of the process temperature, the ambient temperature, the radiation emittance of the surface, the air flow rate over the surface, the k factor, and the thickness of the insulation. The values of R_s encountered vary in the strict sense from 1.71 to 0.25 for hot surfaces and from 2.32 to 0.25 for cold surfaces. These extremes are taken from J. F. Malloy's "Thermal Insulation," Tables 13 and 14. The ASHRAE "Handbook of Fundamentals," however, classified 0.77 to 0.43 as extreme values in Fig. 7 of Chapter 26, reproduced below. This manual recommends the use of 0.7 as the single value to be applied in all cases for solving the economic thickness problem, when more exact values are not readily available. The sensitivity analysis found in Appendix B demonstrates that R_s has a very small influence on the choice of economic thickness. (For the average case, a 50 percent change in R_s results in only a 5 percent change in thickness solution.) This very low sensitivity coupled with the ASHRAE curve, which demonstrates clearly the useful, restricted range of the values, enables the choice of 0.7 to be made for this manual (see Fig. B-4).

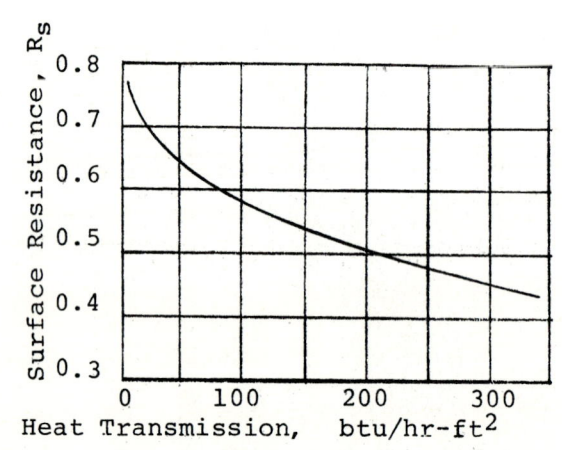

Figure A-2. <u>Heat transmission vs. surface resistance, flat and cylindrical surface</u>

Z

The factor Z is not in itself physically significant, but it carries all of the other variables. At A-20 we let

$$Z = \frac{D}{1.1 m_c B}.$$

1.1, m_c, B and D are analyzed below.

Z Equation Terms

1.1

This multiplier provides 10 percent of annual insulation cost value for the maintenance of the insulation system. Experimental evidence supports this choice of multiplier although extremes of 1.4 to 1.0 are known to exist. Respectively, the range of economic thickness error would be -10 percent to +7 percent if the multiplier were at those extremes.

m_c

The cost rate or incremental cost of applied insulation.

$m_c = \frac{\Delta C}{\Delta L}$, $/in, where ΔC = installed cost difference between two corresponding thicknesses, ΔL. The best data possible should be obtained for these values inasmuch as m_c is the most sensitive parameter in the entire economic thickness solution. If m_c is understated by 50 percent, the insulation thickness chosen will be 50 percent too great; if m_c is overstated by 50 percent, the economic thickness will be 23 percent too small. The chances of being exactly correct in estimating m_c values are small; however, considerable care should be exercized in order to minimize the error in economic thickness (see Fig. B-5).

B3

Any capital investment involves the use of the capital that must be restored and cost of interest on the unrestored capital. Thus, if the investment capital with the interest payments are restored in equal annual increments, we may find those increments by $B = \frac{i_3(1+i_3)^{n_1}}{(1+i_3)^{n_1}-1}$. Whether the capital was obtained from available funds or was obtained by borrowing from the banks makes no difference as long as the money spent is treated as capital and not expense. It is assumed that the insulation project is a capital expense in this manual.

   i_3 The cost of capital as a decimal may be

   --bank rate,
   --opportunity cost,
   --return on investment,
   --or whatever rate the user may want to apply as expressed by the accounting practices of the facility.

An error in this parameter of ±50 percent will produce an average economic thickness error of about -14 percent or +19 percent, respectively (see Fig. B-6).

n_1 The term of years over which the insulation project is to be amortized may be short for "quick payback" conditions or up to 20 years or so for a depreciation or service life type of term. In any case an error in n_1 of ±50 percent will produce an economic thickness of about +18 percent or -23 percent, respectively (see Fig. B-8).

D

As with Z, this term is a mathematic convenience and has little physical significance. However $D_p = -.524 \times 10^{-6} k\ \Delta TYM$ for a round (pipe) system and $D_s = -10^{-6} k\ \Delta TYM$ for a flat surface system of insulation. The parameters k, ΔT, Y, and M will be discussed below.

D Equation Terms

k   This was treated above.

ΔT The difference between the process temperature and the ambient temperature, t_a, is $\Delta T = t_p - t_a$. The parameter is quite sensitive to error-- ±50 percent change in ΔT results in +27 percent, -34 percent change in economic insulation thickness. Hence the choice of t_p and t_a must be made with care (see Fig. B-3).

Y   The number of hours per year that the insulation is expected to conserve heat is Y. If over the years of the insulation project, n_1, the annual utility of the system is expected to change, an average annual time must be chosen. The sensitivity of Y is the same as that for ΔT.

M   The annual average value of energy M, $/10^6$ Btu is given for hot systems, M, or for refrigeration systems, M_r. M and M_r are treated separately.

$$M = 1.1\ AC_h + \frac{P_f B}{Q} \qquad \text{A-34}$$

See Fig. B-9 for the sensitivity of the C_h term.

The terms 1.1, A, C_h, P_f, B, and O are analyzed separately below.

A-34 Equation Terms

1.1

In a given heat-producing facility the cost of heat consists of principally the fuel costs and, secondarily, the other operating and maintenance costs. In 1972, utility O + M costs were 33 percent of fuel costs as reported by the Federal Power Commission. Thus we have

Fuel   1972 = 1.00,   1975 = 2.68
O + M   1972 = .33,   1975 = .50, or 19 percent

of fuel costs in 1975. Fuel costs are rising much faster than O + M costs are rising, so the fuel is becoming a greater percentage of the cost of heat. The choice was then made of O + M being 10 percent of the cost of fuel, as a value that would stand the test of time without causing undue error in the choice of economic thickness. Should, for instance, the 1.1 be truly 1.2, then the economic thickness would be 4 percent too great; should the 1.1 be 1.0 (no O + M costs), the economic thickness would be 3 percent too small.

A

The cost of any good or service which rises at an average rate each year will have an average value over a given span of years. The sum of any series of yearly values which are compounding at a constant rate is given by $\frac{X(1+i)^n - 1}{i}$, which is found in most economic texts as the total amount of an annuity, where

- X is the amount invested per period,
- i is the rate of return per unit time,
- n is the number of time units.

Thus, if we consider the recurring cost of an item to be the first cost compounded each time period (which may be the effect of ordinary inflation), the sum of each of those payments is the same as the total annuity amount above. Now, the average payment per period is merely the sum of all payments divided by the number of periods; therefore,

$$A = \frac{(1+i)^{n_1} - 1}{i_1 n_1}$$

is the average annual payment if the first payment is one (1). A, then is the multiplier for any payment value that is expected to rise at an average rate to yield the average annual payment over the time span of the given project,

    where i_1 is the estimated annual average rate of increase, decimal value (see Fig. B-7).

                   n_1 is the number of years over which the insulation project is being considered. Note that this may be the payback period, the estimated life, the depreciation period or some other time span (see Fig. B-8).

C_h

The cost of 1 million Btu, C_h, $/10^6$ Btu, is the common heat cost relation. C_h is found in any of five ways in this manual:

1. The simple process of buying heat from a commercial source, $/10^6$ Btu.

2. The purchase of electricity from a commercial source, P_e, $/kWh. This is converted to C_h by

$$C_h = \frac{P_e \times 293}{E}, \; \$/10^6 \text{ Btu} \qquad \text{A-35}$$

   where 10^6 Btu = 292.875 kWh (thermochemical calories)

   or = 293.071 kWh (IT calories).

3. The production of heat using fuel oil as the heat source, P_o, $/gal.

$$C_h = \frac{P_o \times 10^6}{w \times E \times H}, \; \$/10^6 \text{ Btu} \qquad \text{A-36}$$

   where P_o = price of oil, $/gal,
   w = specific weight of oil, lb/gal,
   E = conversion efficiency, decimal
   (A boiler may, for instance, have a conversion efficiency of
   $0.8 = \frac{\text{Btu in steam}}{\text{Btu in fuel}}$),
   H = heating value of oil, Btu/lb.

4. The production of heat using coal as the heat source, P_c, $/ton.

$$C_h = \frac{P_c \times 500}{E \times H}, \; \$/10^6 \text{ Btu} \qquad \text{A-37}$$

   where P_c = price of coal, $/ton,
   E = conversion efficiency, decimal,
   H = heating value of coal, Btu/lb.

5. The production of heat using gas as a heat source, P_g, $/1000 ft^3.

$$C_h = \frac{P_g \times 10^3}{E \times H}, \$/10^6 \text{ Btu} \qquad \text{A-38}$$

where P_g = price of 1000 ft^3 gas, \$/1000 ft^3,
E = conversion efficiency, decimal,
H = heating value of gas, Btu/ft^3.

The formulas A-35, A-36, A-37, and A-38 are not derived, as the simplicity of their structure does not warrent the space. If one wants assurance of their accuracy, one can easily check them dimensionally and become confident of their precision.

P_f

The capital investment in a heat producing facility, millions of \$. Thus, if fuel (oil, coal, gas) is burned in a boiler or a powerplant to produce usable heat, the heat cost must carry its share of that capital investment. (See Fig. B-11)

B

See the discussion under Z, preceding.

B in this case is used to find that annual charge for the P_f that must be carried by the Btu's produced. Hence, in the relation

$$B = \frac{i_2(1+i_2)^{n_2}}{(1+i_2)^{n_2}-1},$$

i_2 = facility amortization rate as a decimal (see Fig. B-13),
n_2 = facility amortization life, yrs (see Fig. B-14).

Q

The expected annual Btu production rate, millions of 10^6 Btu/yr, from the heat producing facility is divided into the annual capital charges to give the capital \$/10^6 Btu, which must be added to the fuel costs (see Fig. B-12).

M_r

The annual average value of energy M_r, \$/10^6 Btu for a refrigeration system.

$$M_r = \frac{M}{COP} + 83.3\ SAP_w = 83.3\ \frac{P_r B}{T_r Y} \qquad \text{A-39}$$

The term $\frac{M}{COP}$ is the cost of heat into the system divided by the coefficient of performance which gives the cost of heat removed by the chiller. 83.3 SAP_w adds the average cost of

153

cooling water input to the condenser if the system uses water for the condensing medium. $83.3 \frac{P_r B}{T_r Y}$ provides the annual average amortization of the refrigeration facility per unit of heat removed. Other minor cost elements could be cited but the majority of the costs are covered by this summation. The factors M, COP, S, A, P_w, P_r, B, T_r, and Y are analyzed below.

A-39 Equation Terms

M

The cost of energy discussed above in $/10^6$ Btu is the energy input to the refrigeration system. This usually is in the form of electricity, in kWh, which we convert to 10^6 Btu by multiplying by 293. That times the cost of a kWh gives the energy input cost to the refrigeration system.

COP

COP is an abbreviation for coefficient of performance, which is the ratio of Btu's removed by a refrigeration system to Btu's applied to drive the system.

S

The amount of makeup water required for the condenser in a refrigeration system. Design values are given in Table 3-1. Approximate values can be found as follows:

$$S = 2 \times \frac{(COP + 1)}{COP}$$

This relationship is found from the following:

$G = S(T)$ G is gallons/season,
T is tons refrigeration load,
H is hours used/season,
S is gal/Btu-season.

$$G = T, \text{ton} \times \frac{12{,}000 \text{ Btu}}{\text{ton hr}} \times \frac{\text{lb water}}{1{,}000 \text{ BTU}} \times \frac{\text{gal water}}{8.3 \text{ lb water}} \times H \frac{\text{hr}}{\text{season}} \text{ or } (1.44)TH \frac{\text{gal}}{\text{season}} \text{ for the}$$

refrigeration heat + $\frac{(1.44)}{COP}$ TH for the input heat

$$= (1.44 + \frac{1.44}{COP}) TH = \frac{1.44 \, COP + 1.44}{COP} TH = 1.44 \frac{(COP + 1)}{COP} TH$$

154

without any losses or peripheral heat sources. A safety factor allowance of 1.4 is made which gives

$$S = \frac{1.44 \times 1.4 \ (COP + 1)}{COP} = \frac{2(COP + 1)}{COP}$$

A

In this case, the A is found for the expected rise in the cost of water over the term of the insulation project.

P_w

The price of water, $/gal.

P_r

The capital cost of the refrigeration system or plant that must be amortized over its depreciation life.

B

The amortization factor $\frac{i_2 (1 + i_2)^{n_2}}{(1 + i_2)^{n_2} - 1}$ as previously defined.

T_r

The tons of refrigeration the plant is expected to produce.

Y

The hours per year the plant is expected to function.

APPENDIX B: SENSITIVITY ANALYSIS

The sensitivity of the various parameters to the economic thickness solution was made as follows. This may be termed an error analysis, also.

The solution was expanded to display the user inputs - pipe solution.

$$\frac{(r_2 \ln \frac{r_2}{r_1} + kR_s)^2}{r_2 - kR_s} = \frac{.524 \times 10^{-6} \, k \, \Delta TY}{m_c \left\{ \frac{i_3 (1+i_3)^{n_3}}{(1+i_3)^{n_3}-1} \right\}} \left\{ \left\{ \frac{(1+i_1)^{n_1}-1}{i_1 \, n_1} \right\} \right.$$

$$\left. \frac{1.1C \times 10^6}{EH} + \frac{P}{Q} \left\{ \frac{i_2(1+i_2)^{n_2}}{(1+i_2)^{n_2}-1} \right\} \right\} \qquad \text{B-1}$$

A computer was programmed to solve for r_2 with all elements (17) as variables. Seven different examples of problems were arbitrarily chosen. Each parameter for each problem was independently varied +10 percent, +30 percent and +50 percent. A typical computer output is shown in Fig. B-1. The resulting thickness of insulation chosen for each condition was compared with the no-change or base problem to provide a feeling for the change in thickness with respect to a change in parameter. The percent change in thickness was plotted against the percent change in the parameter. These plots are shown in Figs. B-2 through B-14.

The analysis presented here is for pipe applications. A similar analysis for flat surface applications is not presented since the individual sensitivities for flat surface solutions are generally less than in the pipe solution.

```
EQUATE 1 DATA4
!EQUATE 7 OUTS
!R2MAIN
```

	R2 = 1.6454	+10%	-10%	+50%	-50%
R1	.43000E+00	1.7451	1.5427	2.1217	1.0893
K	.20000E+00	1.7073	1.5305	1.9327	1.2736
RS	.30000E+00	1.6390	1.6517	1.6133	1.6766
DT	.20000E+03	1.7137	1.5741	1.9644	1.2470
Y	.40000E+04	1.7137	1.5741	1.9644	1.2470
VI	.10000E+01	1.5308	1.7211	1.3935	2.2422
I3	.15000E+00	1.6211	1.6707	1.5337	1.7839
N3	.50000E+01	1.6913	1.5947	1.8377	1.3253
I1	.50000E-01	1.6501	1.6407	1.6693	1.6223
N1	.50000E+01	1.6514	1.6394	1.6762	1.6165
C	.50000E-01	1.6913	1.5976	1.3664	1.3900
E	.40000E+00	1.6020	1.6969	1.4802	2.0654
H	.19500E+05	1.6020	1.6969	1.4802	2.0654
P	.10000E+07	1.6679	1.6226	1.7550	1.5230
Q	.20000E+05	1.6247	1.6703	1.5630	1.8535
I2	.70000E-01	1.6639	1.6271	1.7393	1.5580
N2	.40000E+02	1.6416	1.6505	1.6339	1.7029

```
 STOP
 SRU'S: 1.4
 !
```

Figure B-1. Computer readout for sensitivity analysis of various parameters

Figure B-2. Sensitivity analysis of parameter k in equation B-1

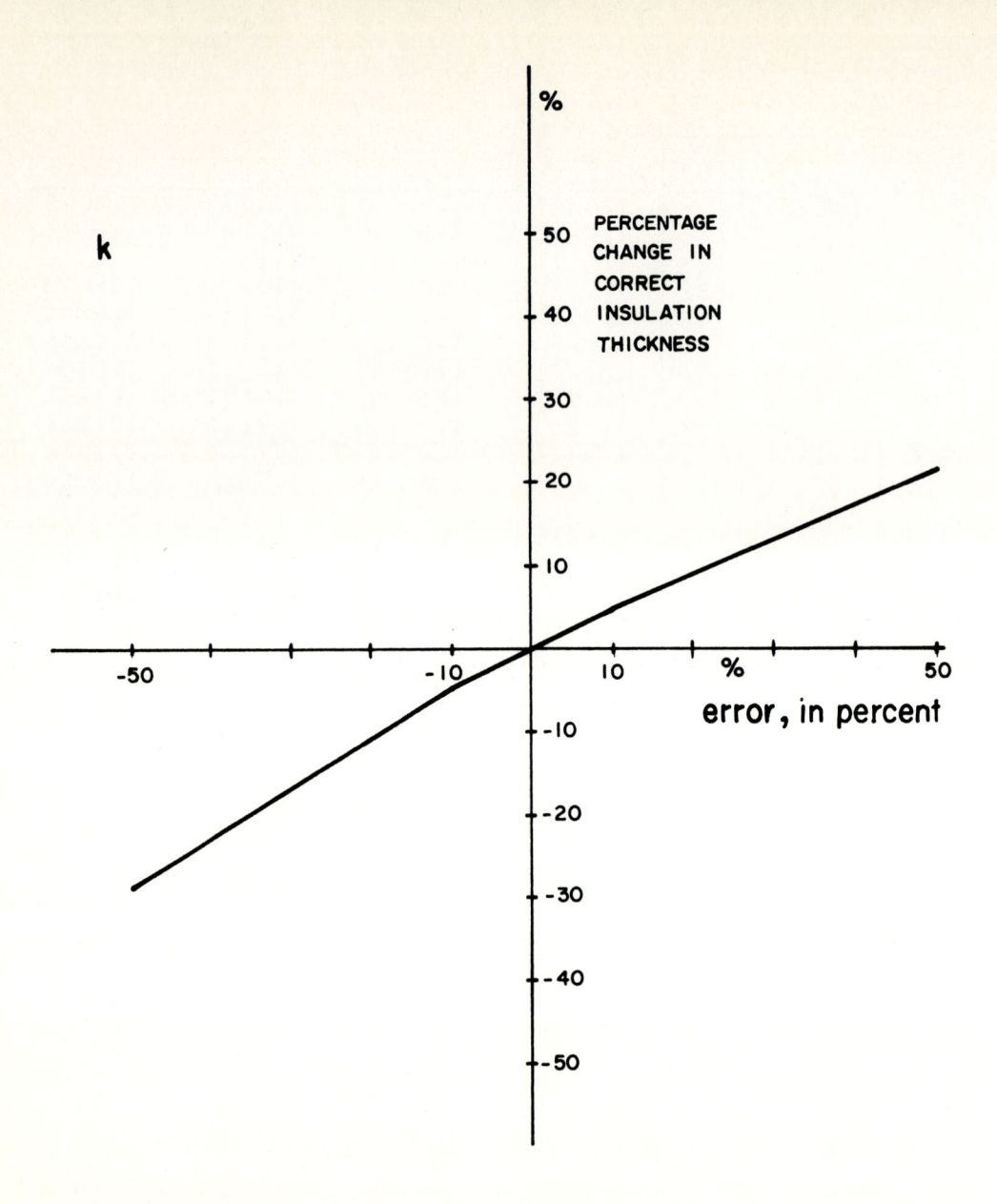

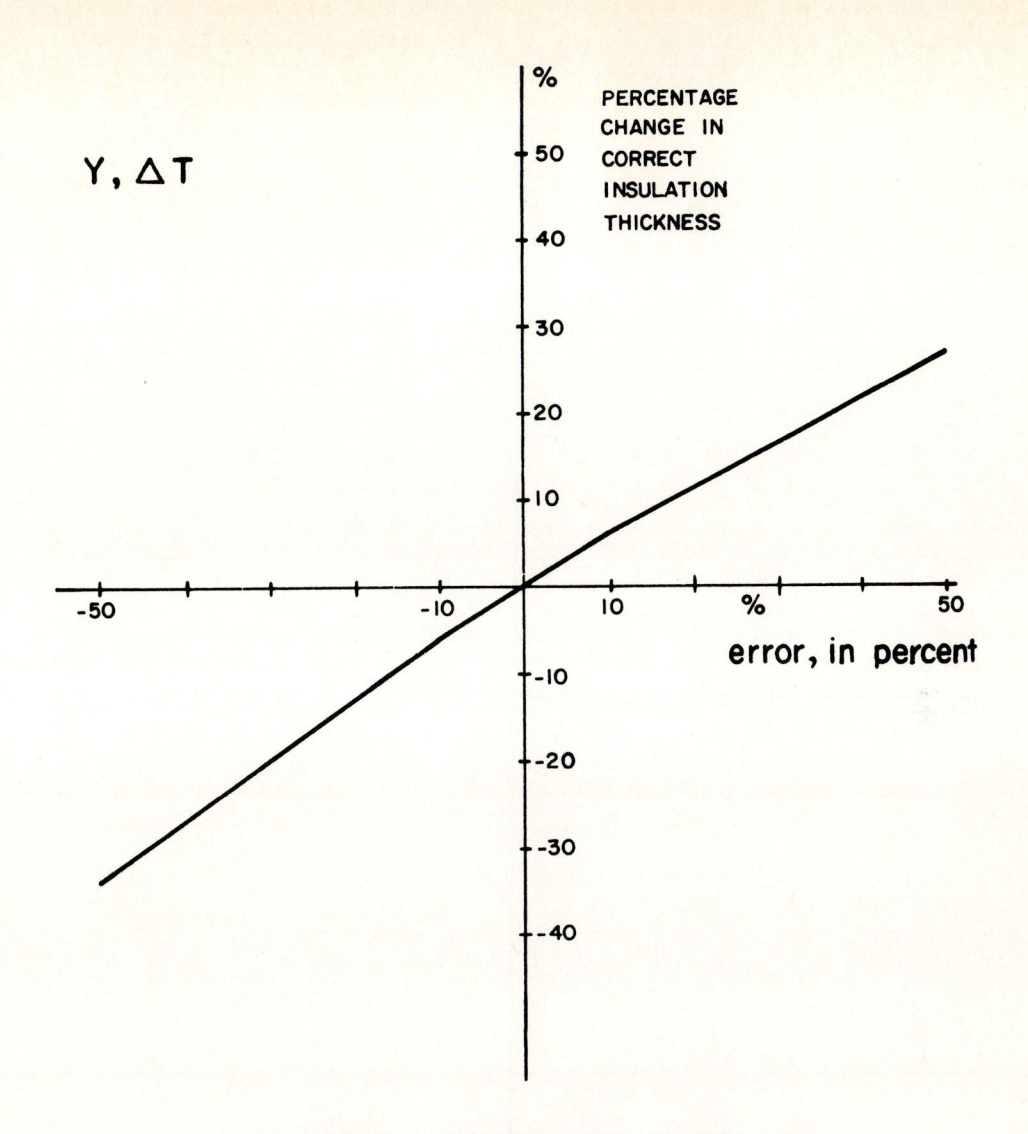

Figure B-3. Sensitivity analysis of parameters Y and ΔT in equation B-1

Figure B-4. Sensitivity analysis of parameter R_S in equation B-1

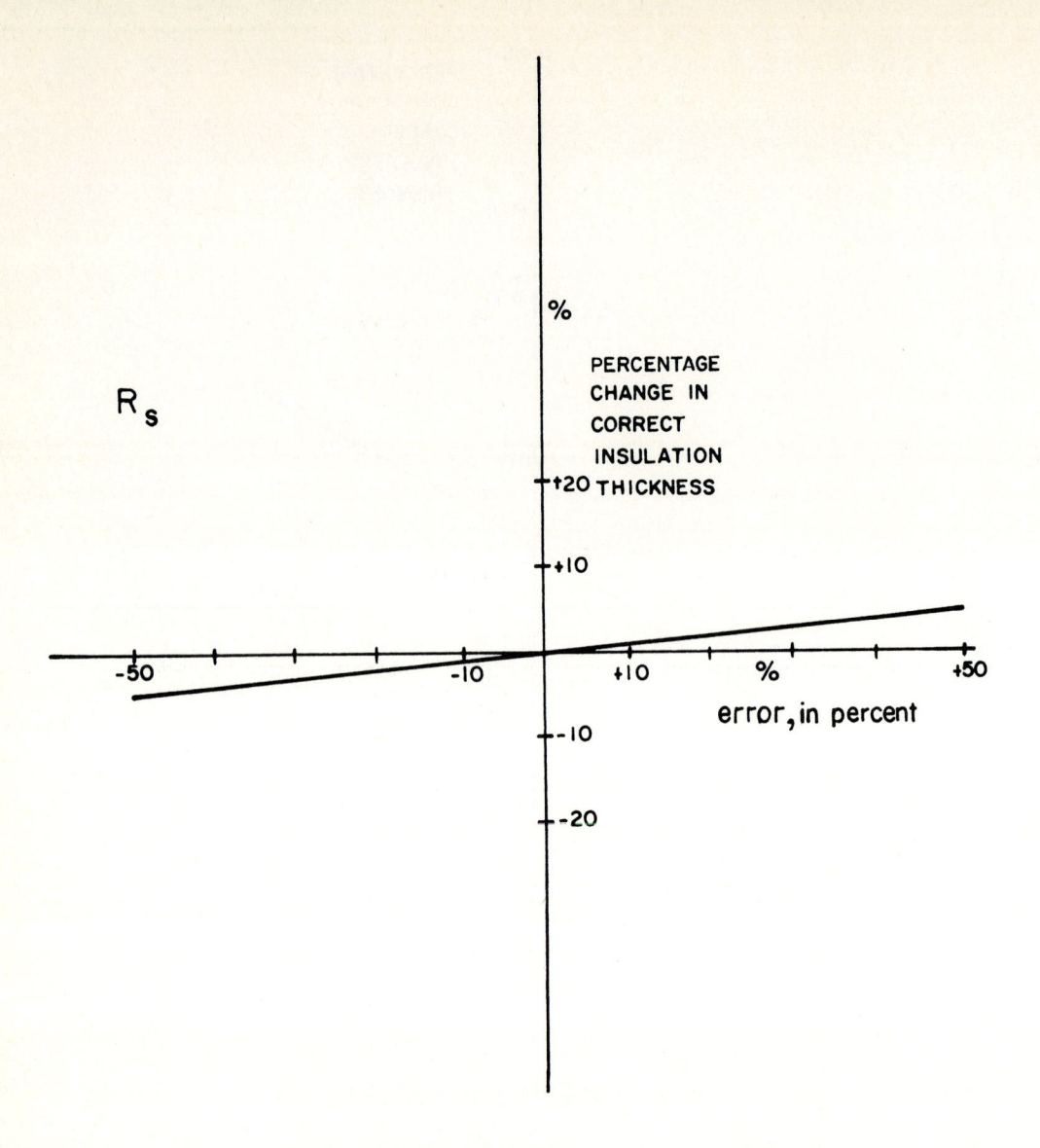

Figure B-5. Sensitivity analysis of parameter m_c in equation B-1

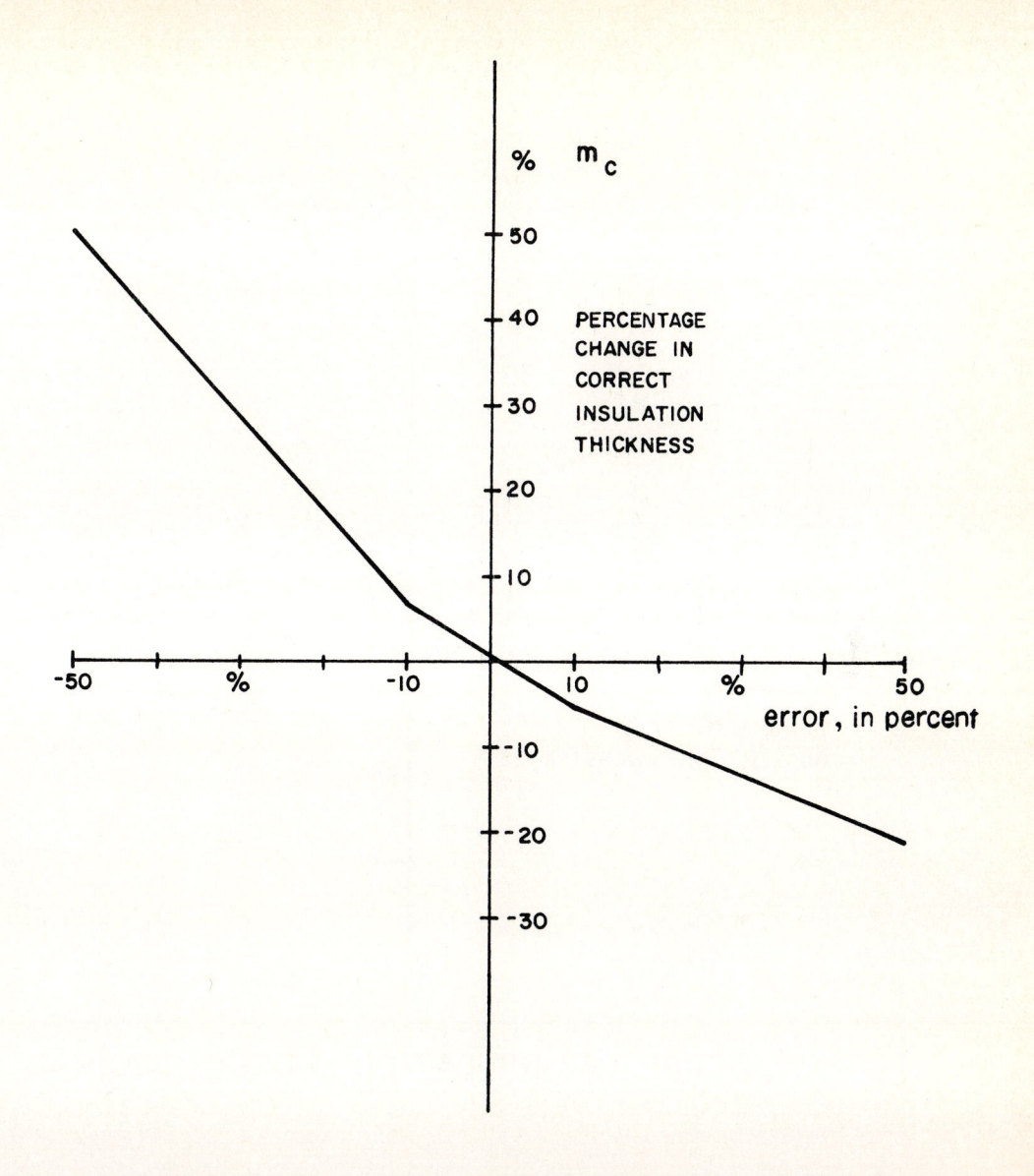

m_c, INCREMENTAL COST OF INSTALLED INSULATION, \$/ft/in

Figure B-6. Sensitivity analysis of parameter i3 in equation B-1

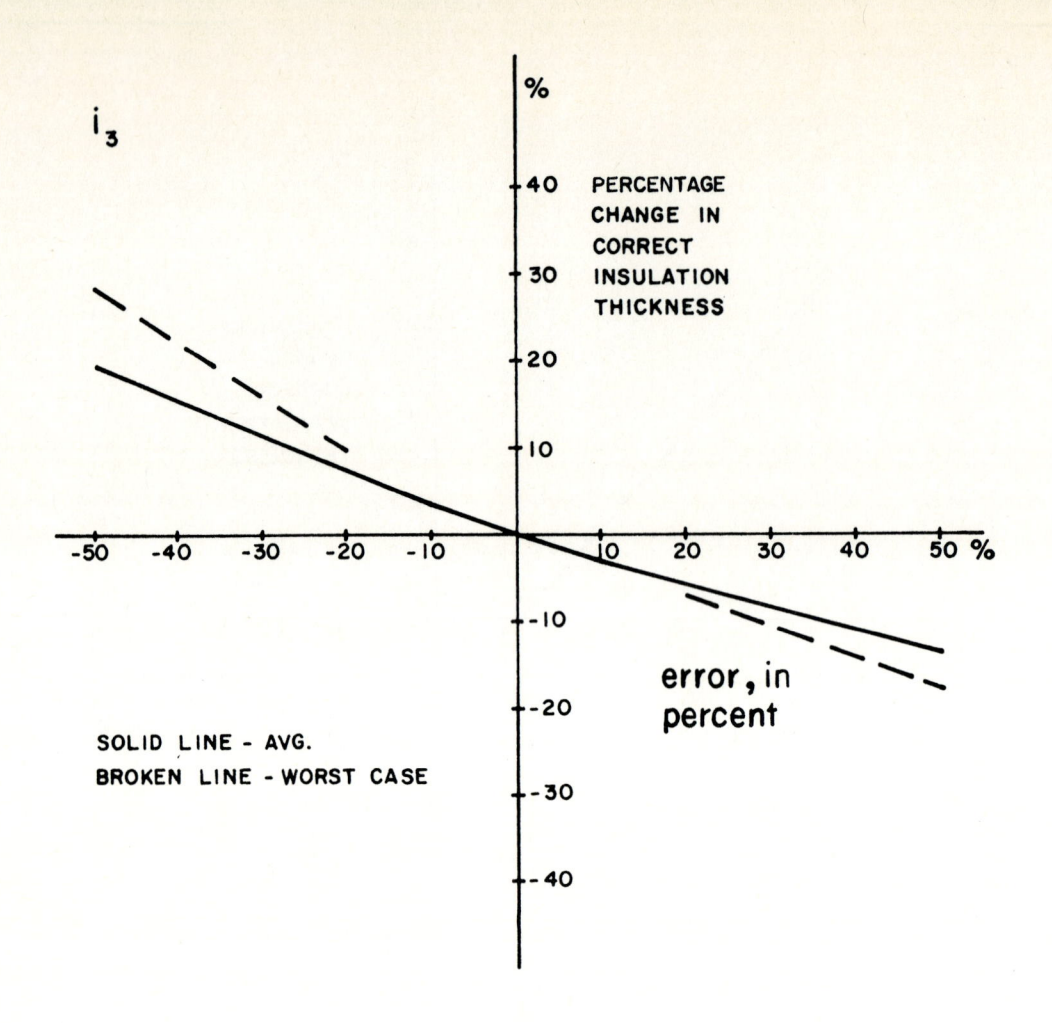

Figure B-7. Sensitivity analysis of parameter i_1 in equation B-1

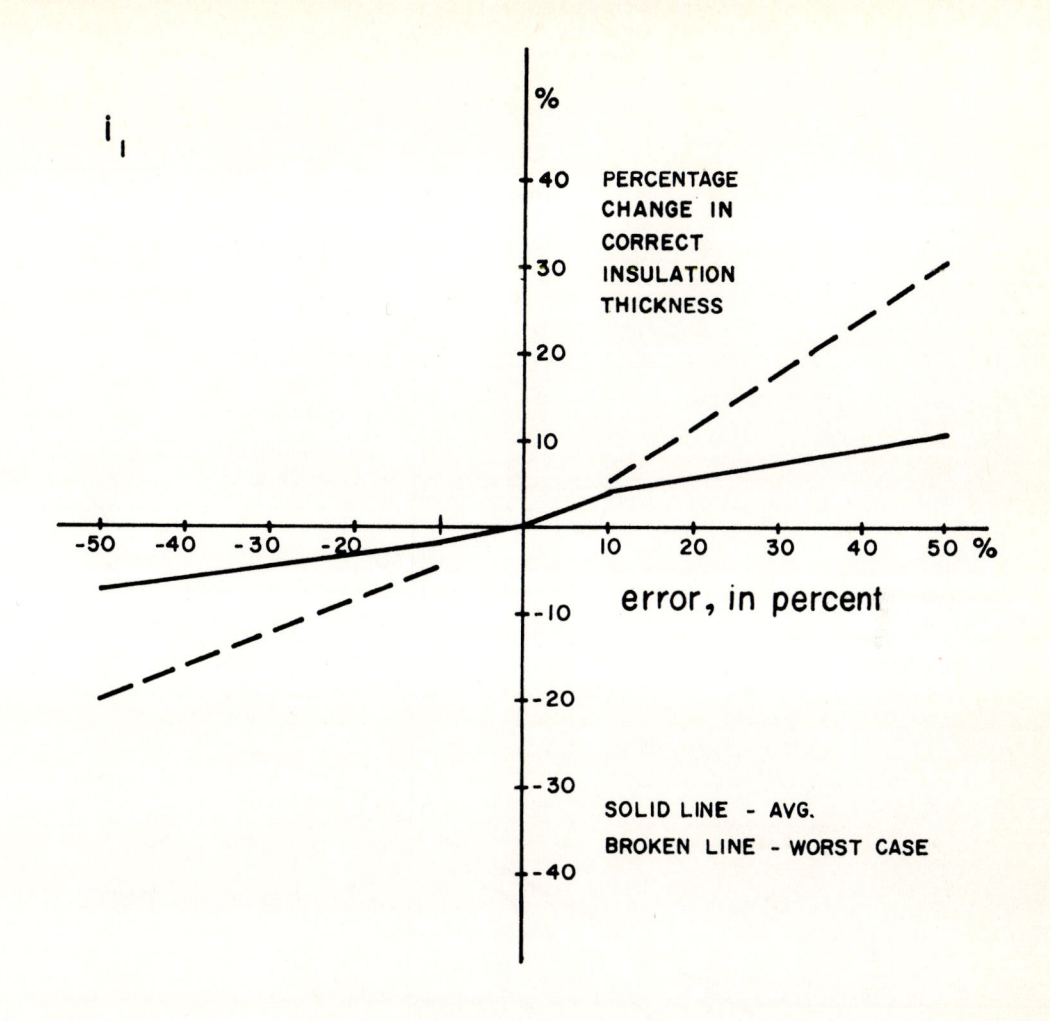

Figure B-8. Sensitivity analysis of parameter n_1 in equation B-1

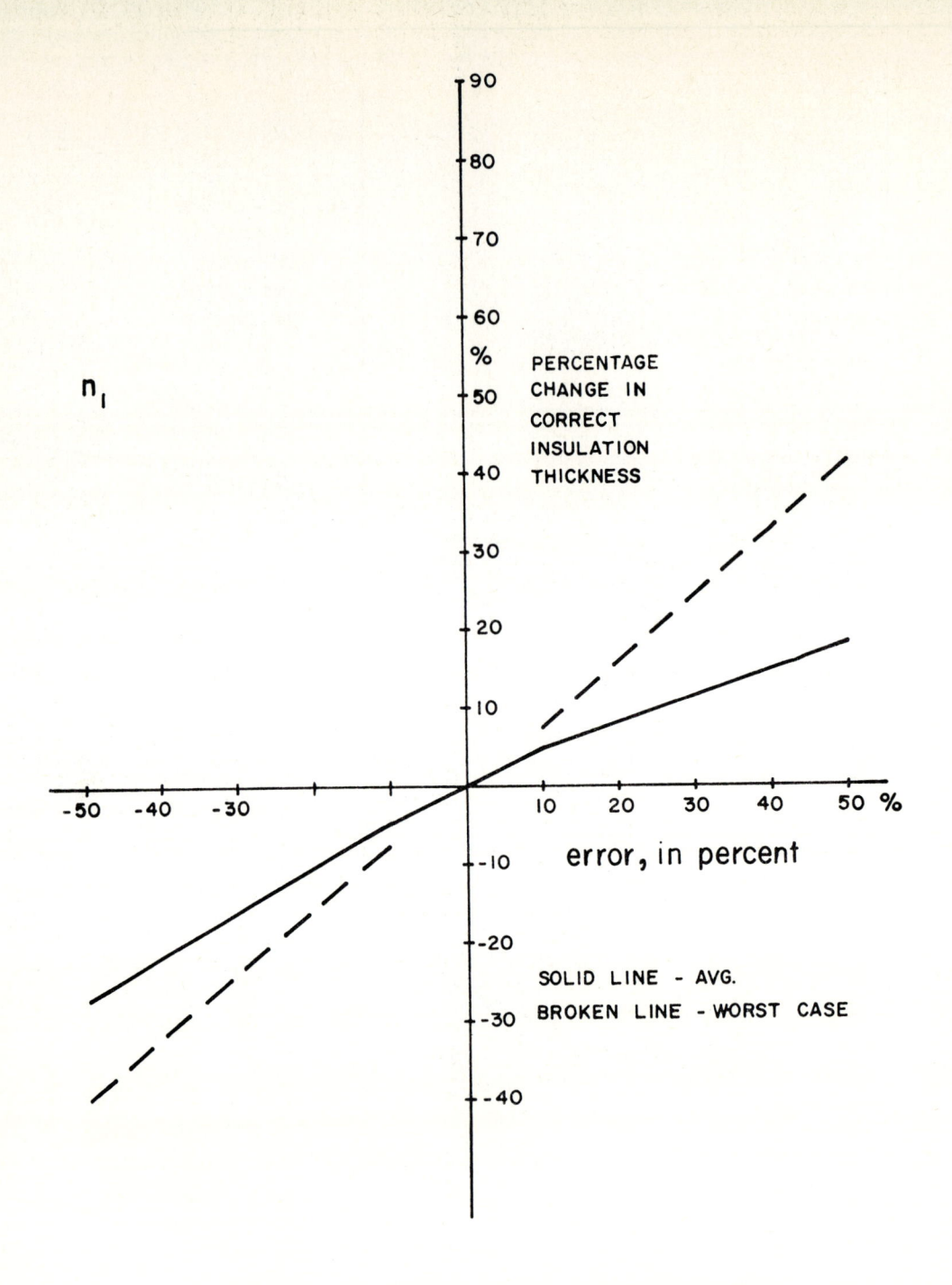

Figure B-9. Sensitivity analysis of parameter C in equation B-1

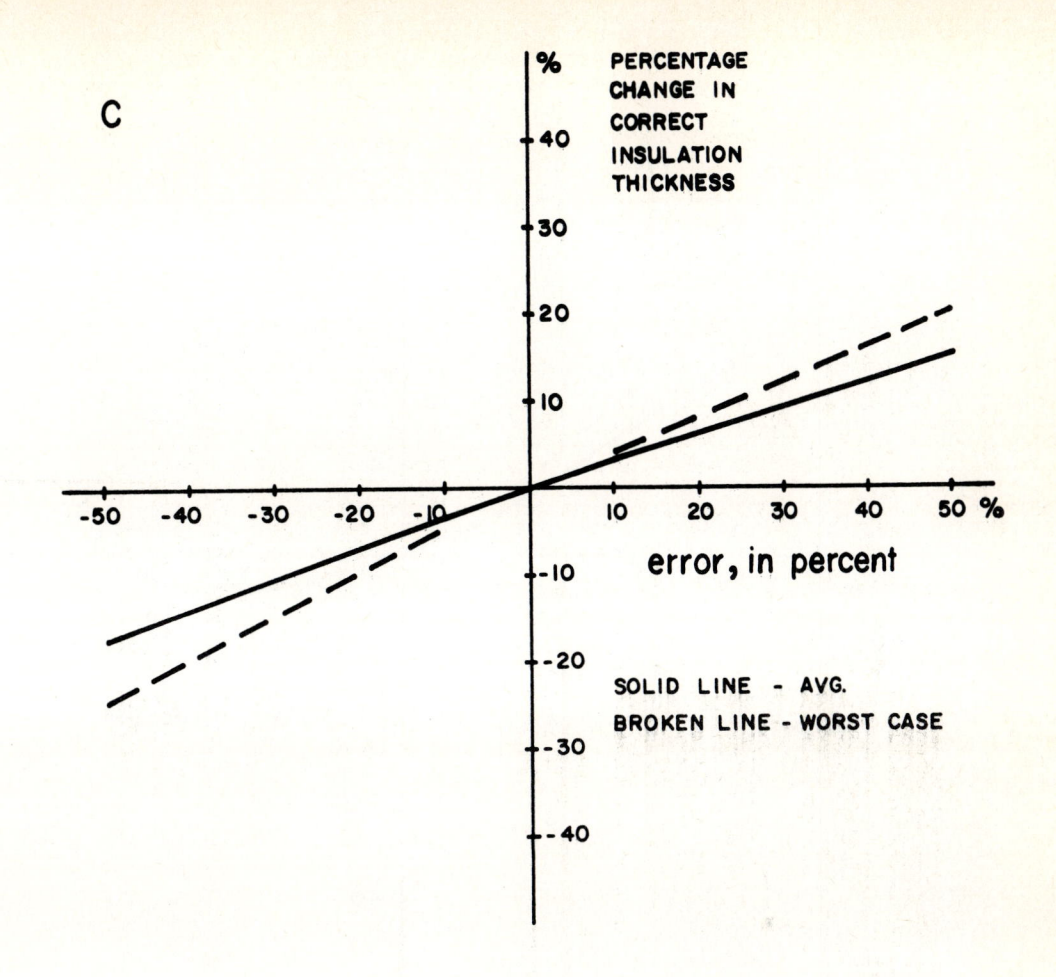

Figure B-10. Sensitivity analysis of parameters E and H in equation B-1

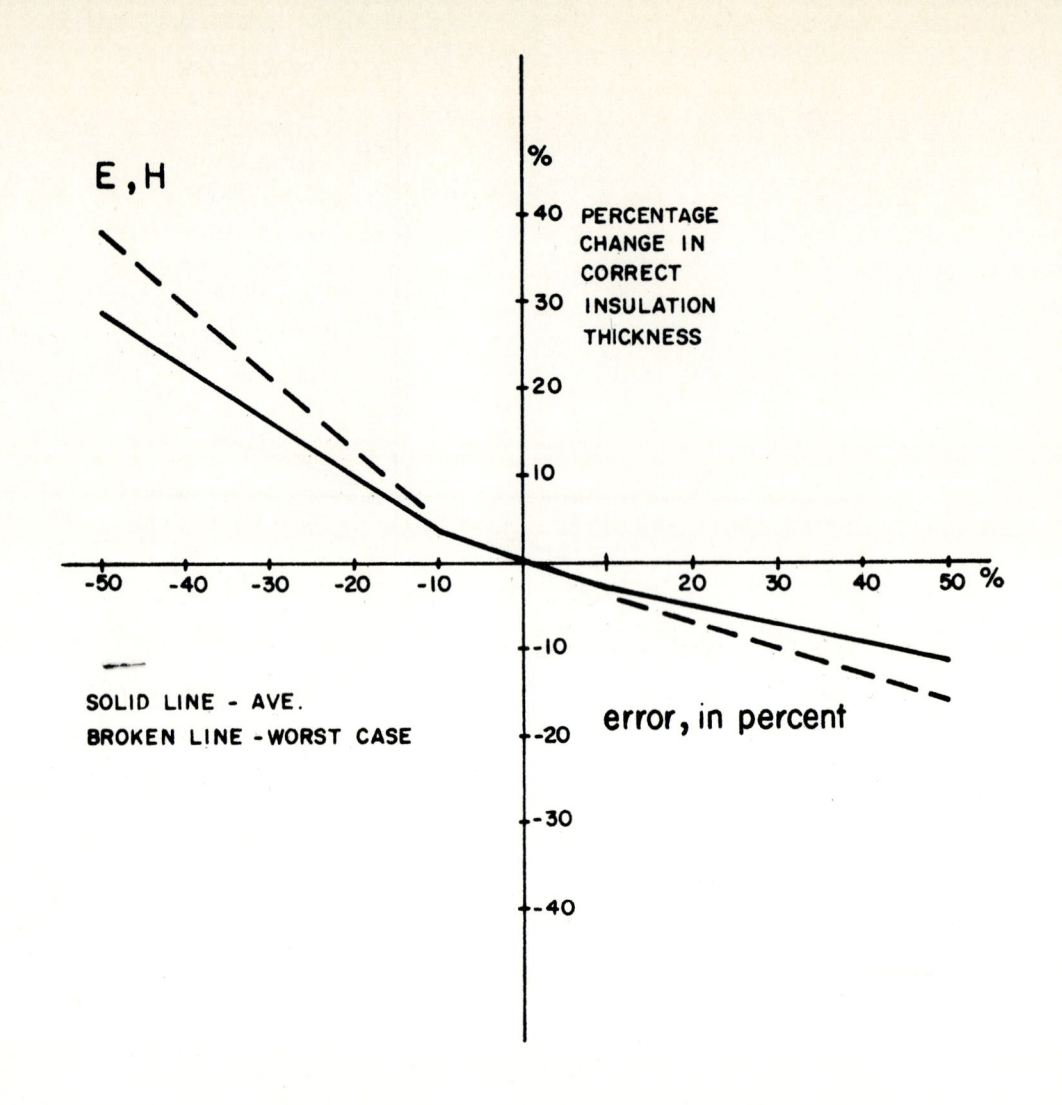

Figure B-11. Sensitivity analysis of parameter P in equation B-1

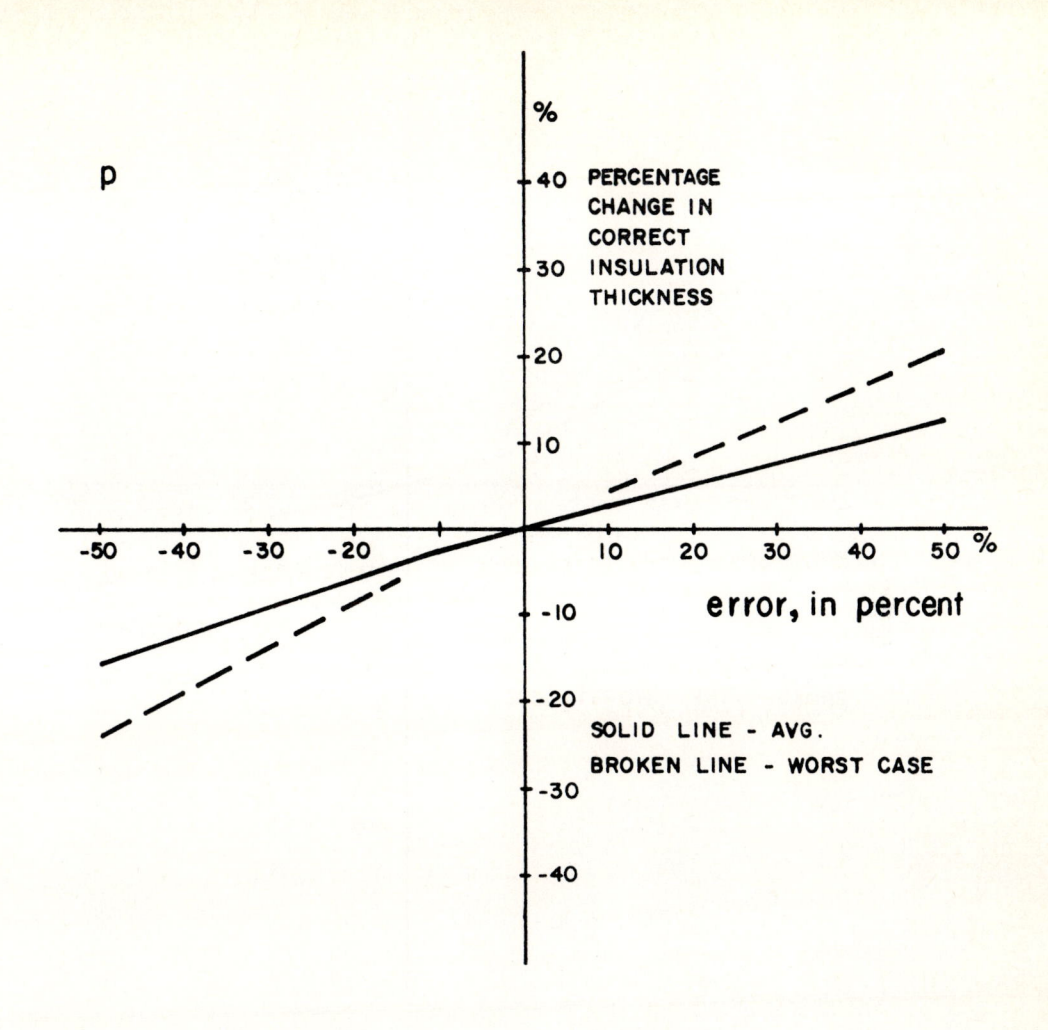

p, CAPITAL COST OF HEAT PRODUCTION FACILITY, $

PERCENTAGE CHANGE IN CORRECT INSULATION THICKNESS

error, in percent

SOLID LINE - AVG.
BROKEN LINE - WORST CASE

Figure B-12. Sensitivity analysis of parameter Q in equation B-1

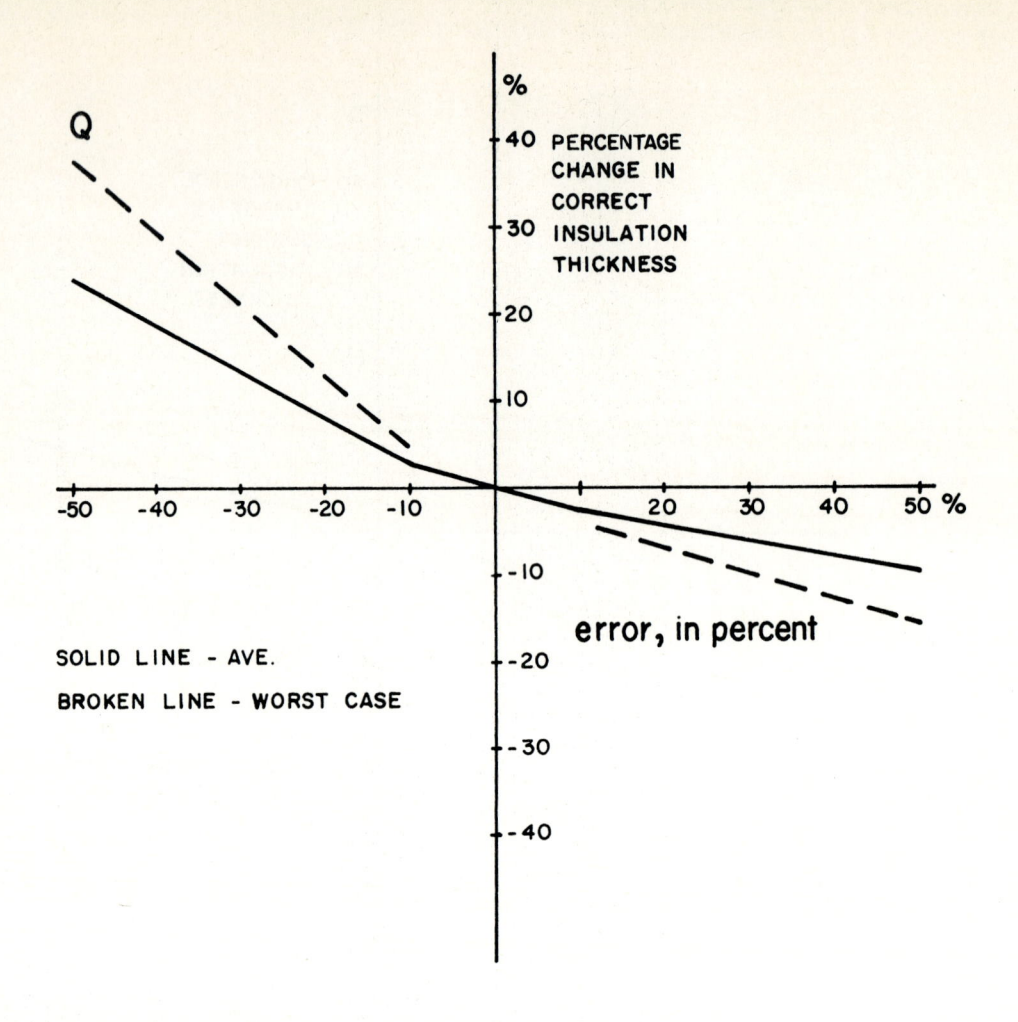

Figure B-13. Sensitivity analysis of parameter i_2 in equation B-1

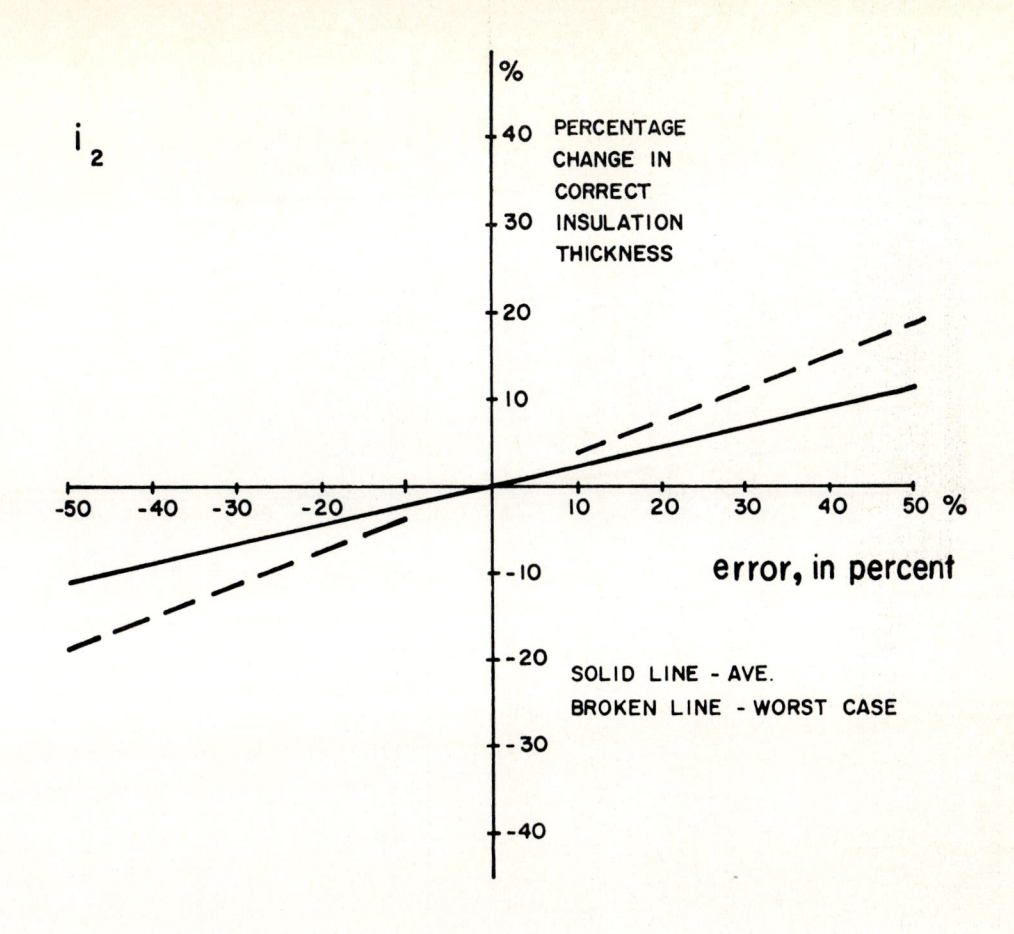

Figure B-14. Sensitivity analysis of parameter n_2 in equation B-1

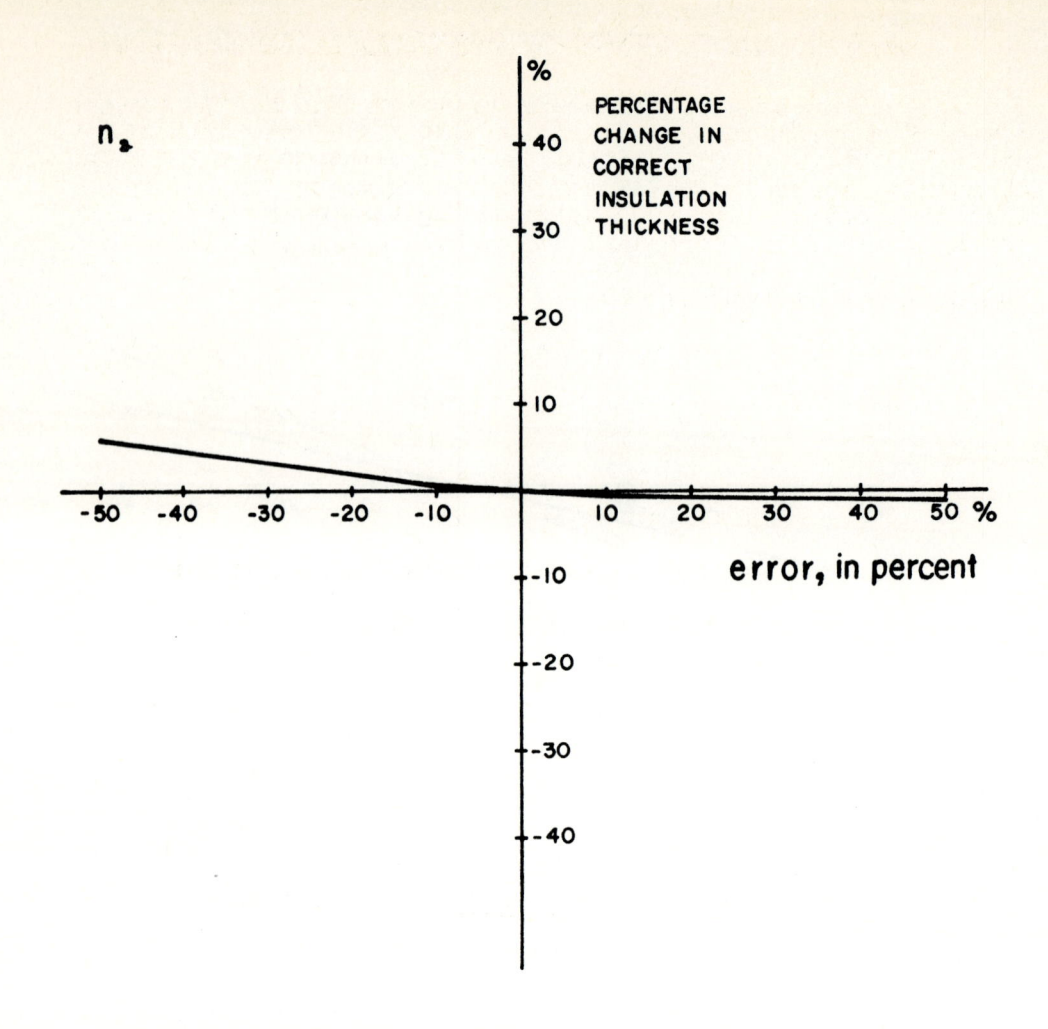

APPENDIX C: COST OF INSTALLED INSULATION

The cost of installed insulation is determined by the simple sum of the variable and the fixed portions of the insulation project. The variable portion is a function of the thickness of the insulation. Therefore, unit consistency is maintained with the cost of lost heat which has utilized the insulation thickness as its independent variable.

The cost may be expressed as

$$P_i = m_c L + d \qquad \text{C-1}$$

where P_i is the cost of the installed insulation, $/ft^2$

m_c is the rate of installed insulation cost, $/ft^2$/in

L is the thickness of the insulation, in

d is the fixed (non-variable) installation cost, $/ft^2$

This has the appearance of a straight line, but m_c may not be a constant. For instance, data used by one large firm was found to fit $P_i = .5L^2 + .16 L + 4$, so that $\dfrac{dP_i}{dL} = m_c = L + .16$.

Other data has been shown to have the form $m_c = \dfrac{ab(L+c)^b}{L+c}$, where a, b, and c are constants.

There is little evidence that a single consistently accurate model can be established because of the kind and variety of variables that enter the cost, P_i. Some of the cost elements that enter this installed insulation factor are as follows:

- A   Labor cost and productivity
- B   Union vs. non-union labor
- C   Transportation costs
- D   Complexity of the system--number of valves, elbows, and fittings in the system
- E   Accessibility of the installation
- F   Large or small job
- G   Kind of insulation
- H   Form of insulation
- I   Profit margins
- J   Overhead margins
- K   Overtime

L   Trade discounts
M   Geographic location of job
N   Seasonal and weather considerations.

Inasmuch as many variables exist and we cannot predict those to be encountered by users of this manual in solving specific problems, it is a requirement that we find a model accurate enough to provide least cost insulation solutions. A further complication is that the sensitivity analysis shows clearly that an error in m_c is the most troublesome of all parameter errors. A 50 percent error of m_c on the low side will result in an insulation thickness which is 50 percent greater than the optimum thickness, while a 50 percent high error for m_c will result in insulation thickness which is 23 percent less than optimum. Fortunately, an estimate for m_c on the low side will result in an insulation thickness on the thick side, which will save energy if not dollars.

By evaluating a series of costs from widely diverse sources a compromise solution is found that yields high levels of accuracy without excessive effort. Figure C-1 shows real data plotted for a commonly occurring installation case. The plotted data reveal that a smooth curve could be used to match each set of points. It is also apparent that a common curve is not accurate enough to reach the least cost insulation thickness solution. Two linear fits are made in the single layer areas:

1. The solid line merely connects the one-inch and three-inch data.
2. The dotted line is least squares fit of the single layer data. That these two estimates of the slope P_i vs. L are so similar is the source of the final choice of solution for m_c as used in this manual.

We chose m_c by taking two separate installed insulation estimates in the single layer thickness range, and then took the difference of the P_i values, and divided by the difference of the L values, with m_{c1} the single layer slope.

$$m_{c1} = PC \frac{P_{i2} - P_{i1}}{L_2 - L_1}$$

where PC is the job complexity factor used to account for fittings in the system (See Fig. C-2).

We chose two thicknesses in the double layer range, found the respective installed insulation costs, took the difference between the P_i values, and divided by the difference between the L values to get m_{c2}, the double layer slope.

$$m_{c2} = PC \frac{P_{i4} - P_{i3}}{L_4 - L_3} , \text{ (see Fig. C-2).}$$

Likewise for the triple layer range we have $m_{c_3} = PC \dfrac{P_{i_6} - P_{i_5}}{L_6 - L_5}$ (see Fig. C-2).

Having determined m_{c_1}, m_{c_2}, and m_{c_3}, the relation $P_i = m_c L + d$ may be used with a high degree of reliability since the m_c is broken into small enough fragments to retain its accuracy within the range for which it was determined.

Note that the fixed cost term, d, was not evaluated since only the variable costs are needed in the solution (see A-17).

The piping complexity factor, PC, as explained in Appendix F, is finally multiplied to complete the m_c term

$$m_c = PC \frac{\Delta P}{\Delta L}$$

Maintenance Costs

Insulation has to be given some maintenance. Maintenance expense was chosen to be 10 percent of the original cost spread (prorated) over the life-time of the insulation project (n_1). Evidence of a 5 to 40 percent maintenance cost is available. This gives a 10 percent mean value +30 percent-5 percent for the data. Using the sensitivity analysis, we show insulation thickness will vary -14 percent to +5 percent as a result of maintenance costs being other than the 10 percent mean value used in Equations A-14 or A-31.

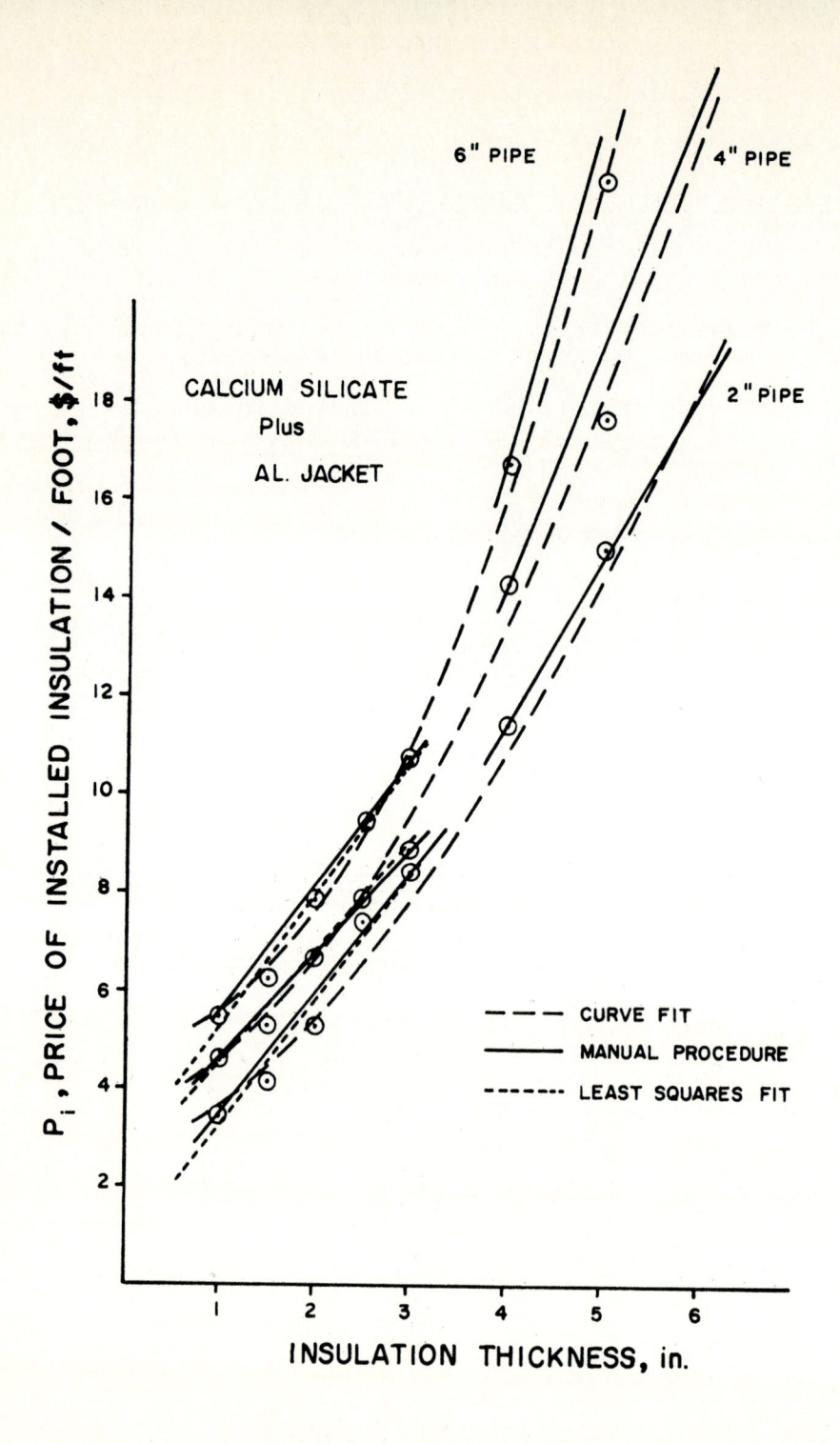

Figure C-1. Plot of actual data with various means of fitting data points

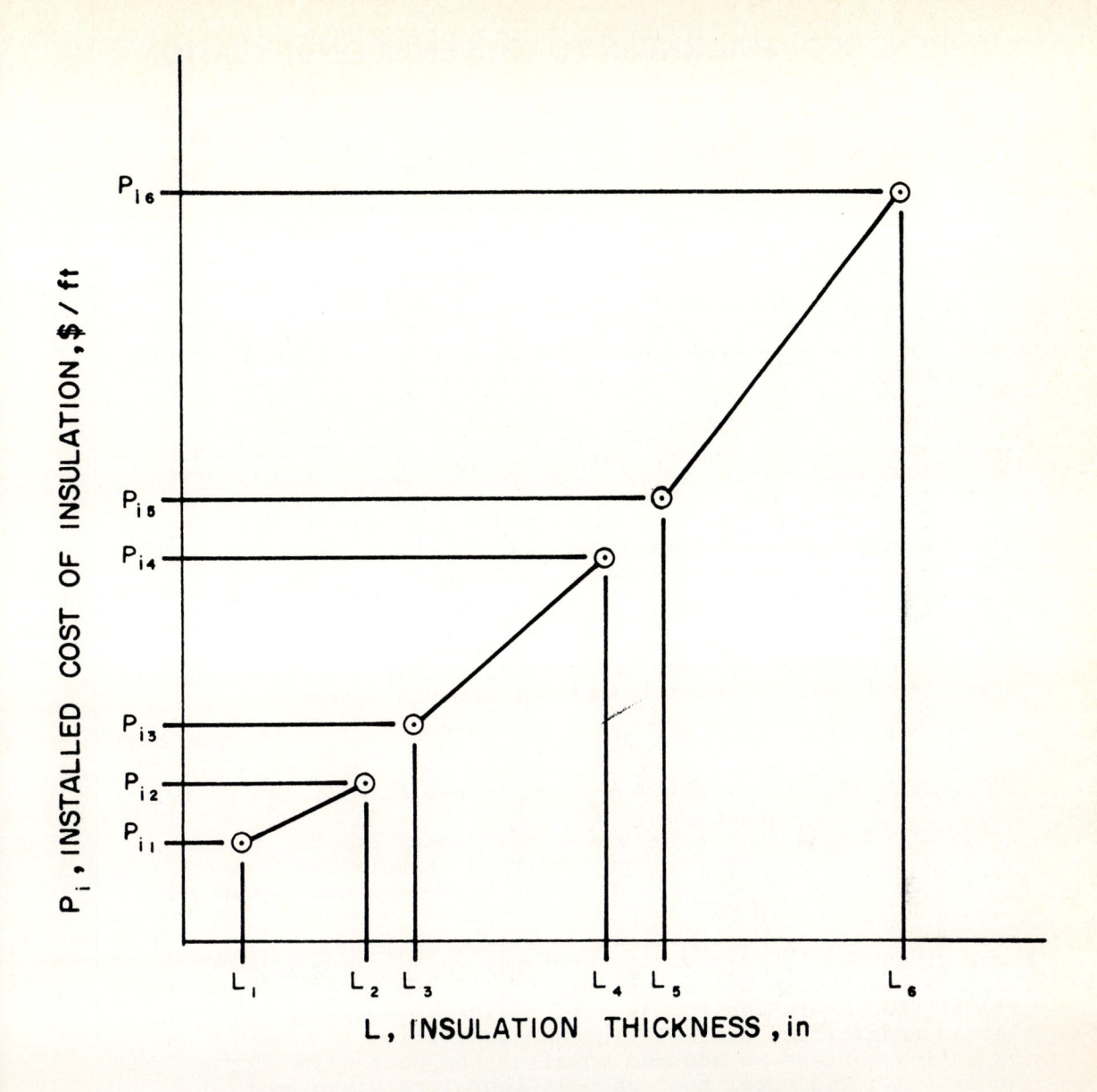

Figure C-2. Data requirements for m_C determination

APPENDIX D: THICKNESS TO PREVENT CONDENSATION

Condensation on the outside surface of insulation can be prevented if there is sufficient thickness to keep the surface temperature above the dewpoint.

The dewpoint temperature is determined from the relation

$rh = \dfrac{e''_w}{e_w}$ where rh is the relative humidity, e''_w is the vapor pressure corresponding to the dewpoint temperature, and e_w is the vapor pressure corresponding to the dry bulb temperature. The nomograph for the dewpoint temperature synthesizes that relation by superimposing the temperatures on a vapor pressure solution. Thus, it is necessary to find the relative humidity from a wet-dry dry bulb chart and set the dry bulb and relative humidities on the nomograph to get t_d, the dewpoint temperatures. The values obtained are those at standard sea-level pressures which is normally not further corrected for altitude. For instance, using a dry bulb of 100°F and a wet bulb of 81°F the relative humidity at sea level is found to be 45 percent; however, at 5,000 feet the relative humidity would be only 46 percent. This represents a dewpoint change of 1/2°F, which is a trivial change.

The design surface temperature of the insulation corresponding to the local design ambient air conditions is given as

$$t_s = t_d + 1 \qquad \text{D-1}$$

where t_s = the design surface temperature of the insulation, in °F
t_d = the design dewpoint temperature for a given design condition of relative humidity and dry bulb temperature expressed in °F.

For a given design surface temperature, the minimum insulation thickness is determined by calculating the rate of heat transfer through the insulation to the cold source. The amount of heat that is transmitted by radiation and convection to the insulation surface is assumed equal to the heat gain of the insulation. The total heat gain is therefore given as

$$Q_a = Q_r + Q_{cv} \qquad \text{D-2}$$

where Q_a = the total heat gain to the insulation in $\dfrac{Btu}{hr\text{-}ft^2}$
Q_r = the heat absorbed by the insulation surface due to the radiation from the higher temperature ambient air to the lower temperature insulation surface (Btu/hr-ft^2).

Q_{cv} = the heat absorbed by the insulation surface from convective air currents, Btu/hr-ft^2.

Radiative heat gain, according to the Stefan-Boltzman Law, is given as

$$Q_r = .174\varepsilon \left[\left\{\frac{T_a}{100}\right\}^4 - \left\{\frac{T_s}{100}\right\}^4 \right] \qquad \text{D-3}$$

where T_s = the insulation surface temperature, °R
($°R = 460 + °F$)
T_a = the design ambient dry bulb temperature, °R
ε = the surface emittance of the insulation finish or jacket, expressed as a decimal.

The emittance of the surface is the measure of its ability to emit radiant energy relative to a black body. It varies with the type, color, and condition of the surface. Surfaces with high emittance will absorb more heat from the surrounding air than will surfaces of low emittance. As a result, surfaces of high emittance maintain a higher insulation surface temperature and require less of an insulation thickness to prevent condensation. Therefore, a black surface ($\varepsilon = 1.0$) is the best for minimizing insulation thickness, to prevent condensate formation.

Heat gain from convection, according to Langmuir's equation, is given as

$$Q_{cv} = .296 (t_a - t_s)^{1.25} \qquad \text{D-4}$$

where t_s = the surface temperature of the insulation, in °F
t_a = the design ambient dry bulb temperature, in °F.

The above equation assumes convection from natural air circulation, which is the worse design condition. Forced circulation minimizes the possibility of moisture condensing on cold surfaces.

The total heat transmitted to the insulation surface (Q_a) is used to determine the minimum insulation thickness.

$$Q_a = \frac{k(t_s - t_p)}{L} \qquad \text{D-5}$$

where $Q_a = Q_r + Q_{cv}$, in $\frac{Btu}{hr-ft^2}$

k = the thermal conductivity of the insulation in $\frac{Btu-in}{hr-ft-°F}$

t_p = the process temperature of the cold source to be insulated, °F

t_s = the design surface temperature of the insulation, °F

L = the equivalent thickness of insulation, in inches.

Combining equation 3, 4, and 5 results in D-6, which is used to solve for equivalent insulation thickness to prevent condensation. Conversion of equivalent thickness to nominal thickness for various pipes is accomplished by Figure 8-1 in Chapter 6.

$$L = \frac{k(t_s - t_p)}{.174\varepsilon\left[\{\frac{T_a}{100}\}^4 - \{\frac{T_s}{100}\}^4\right] + .296(t_a-t_s)^{1.25}} \qquad \text{D-6}$$

Equivalent thickness is the thickness of insulation on a flat surface which would be required to give the same rate of heat transmission per square foot of outer surface insulation as on a cylinder or pipe.

$$L = r_2 \ln \frac{r_2}{r_1}, \text{ in} \qquad \text{D-7}$$

where r_2 is the insulation outer radius, in
r_1 is the insulation inner radius, in

Parameter Sensitivity

The insulation thickness required to prevent condensation is given by

$$L = \frac{k(t_s-t_p)}{.174\varepsilon\left[\{\frac{T_a}{100}\}^4 - \{\frac{T_s}{100}\}^4\right] + .296(t_a-t_s)^{1.25}}$$

where k is the insulation thermal conductivity coefficient, $\frac{Btu-in}{°F-hr-ft^2}$

t_s is the required surface temperature, °F
t_p is the process temperature, °F
ε is the emittance in calm air of the outer surface, decimal
T_a is the ambient temperature, °R
T_s is the surface temperature, °R
t_a is the ambient temperature, °F

A complete study was made of the effect of various errors on the insulation thickness required to prevent condensation. The results of this study are shown graphically in Fig. D-1. The proper choice of dewpoint temperature is obviously the most critical item in the solution. This sensitivity is so great that care must be used at each step to enter the proper value, or a significant departure from the appropriate insulation thickness will occur.

A sample computer data sheet is provided as Figure D-2.

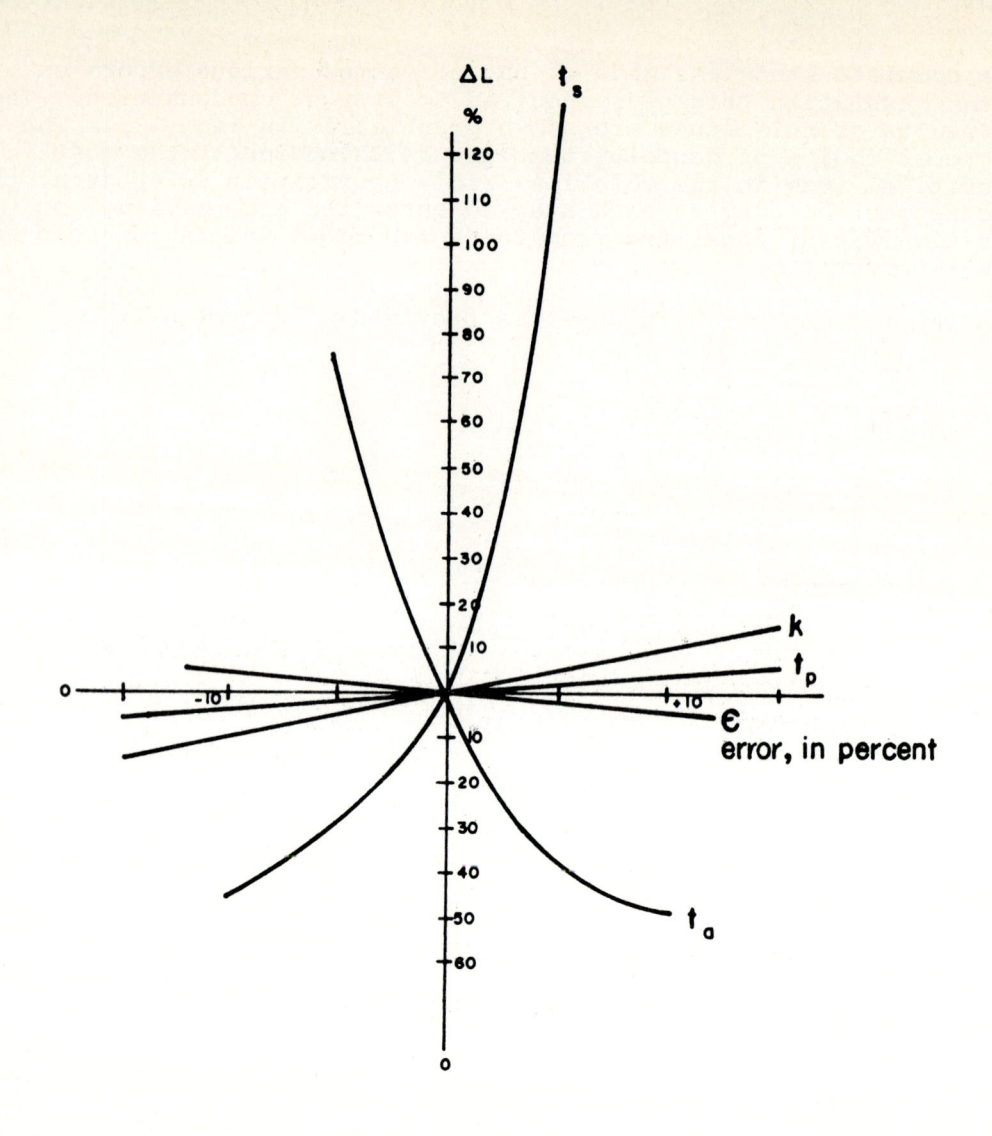

Figure D-1. Parameter sensitivity for L, insulation thickness to prevent condensation

180

```
* 
LMAIN
     L =      .4567
```

			L	DELTA L	% DIFF
K		.12000E+00			
	+10%	.13200E+00	.5023	.0457	10.0000
	-10%	.10800E+00	.4110	.0457	10.0000
	+50%	.18000E+00	.6850	.2283	50.0000
	-50%	.60000E-01	.2283	.2283	50.0000
TS		.40000E+02			
	+10%	.44000E+02	.6700	.2134	46.7231
	-10%	.36000E+02	.3363	.1204	26.3624
	+50%	.60000E+02	-5.6401	6.0968	********
	-50%	.20000E+02	.1379	.3188	69.8121
TP		-.40000E+02			
	+10%	-.44000E+02	.4795	.0228	5.0000
	-10%	-.36000E+02	.4338	.0228	5.0000
	+50%	-.60000E+02	.5708	.1142	25.0000
	-50%	-.20000E+02	.3425	.1142	25.0000
E		.90000E+00			
	+10%	.99000E+00	.4315	.0252	5.5209
	-10%	.81000E+00	.4850	.0283	6.2061
	+50%	.13500E+01	.3534	.1033	22.6110
	-50%	.45000E+00	.6452	.1885	41.2776
TA		.55000E+02			
	+10%	.60500E+02	.3202	.1364	29.8742
	-10%	.49500E+02	.7625	.3058	66.9741
	+50%	.82500E+02	.1374	.3193	69.9196
	-50%	.27500E+02	-3.8875	4.3442	951.2906

```
STOP
SRU'S:.6
!
```

Figure D-2. Sample computer run of condensation control sensitivity analysis

APPENDIX E: RETROFIT – MULTILAYER INSULATION

Pipe

The quantity of heat transfered to or from a pipe with two layers of insulation with dissimilar thermal conductivities is closely approximated as follows:

$$\frac{q_p}{A_3} = \frac{\Delta T}{\frac{r_3}{k_1} \ln \frac{r_2}{k_1} + \frac{r_3}{k_2} \ln \frac{r_3}{r_2} + R_s} \qquad \text{E-1*}$$

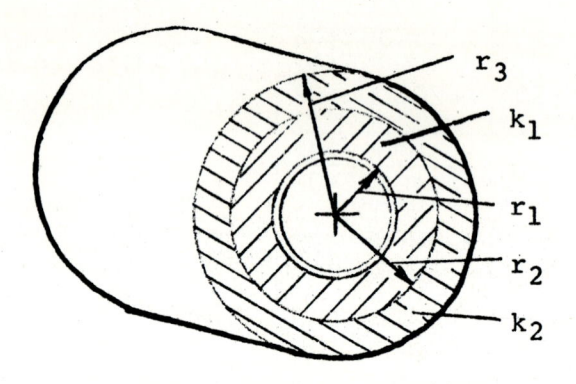

Figure E-1. Pipe with two insulation layers

where q_p heat transfer from temperature t_p to temperature t_a, $\frac{Btu}{hr}$

A_3 is outer area, ft^2
r_1 is pipe radius, in
r_2 is 1st insulation layer radius, in
r_3 is 2nd insulation layer radius, in
k_1 is 1st layer coefficient, $\frac{Btu\text{-}in}{°F\text{-}hr\text{-}ft^2}$
k_2 is 2nd layer coefficient, $\frac{Btu\text{-}in}{°F\text{-}hr\text{-}ft^2}$
R_s is surface resistivity.

*See Malloy, J. F. op.cit. p. 12, eq, (10).

Rearrange

$$\frac{q_p}{A_3} = \frac{k_1 k_2 \Delta T}{k_2 r_3 \ln \frac{r_2}{r_1} + k_1 r_3 \ln \frac{r_3}{r_1} + k_1 k_2 R_s}$$

Solve for heat flow per linear foot of pipe

$$U_p = \frac{.524 \, k_1 k_2 \, \Delta T \, r_3}{k_2 r_3 \ln \frac{r_2}{r_1} + k_1 r_3 \ln \frac{r_3}{r_2} + k_1 k_2 R_s}$$

$$\frac{Btu}{hr\text{-}ft} \qquad \qquad \text{E-2}$$

Assumptions:

1. k_1 and k_2 are chosen from mean temperatures which are found by assuming that the insulation layers equally divide the regime t_p through t_a. This simplification will seldom be exact, but the rigor of finding the true (?) k values by iteratively solving the problem is not needed for the least cost insulation thickness choice. However, if the solution for outer surface temperature was being made, then an additional effort would be necessary and proper. (True k values are not available from any source at this time, but values which are useful may be obtained from design curves and data.)
2. The temperature drop across the pipe is neglected.

Annual cost of lost heat is

$$m_p = U_p \, YM \times 10^{-6}, \; \$/yr\text{-}ft \qquad \qquad \text{E-3}$$

where U_p is given by eq. E-2
          Y is annual operating time, hr
          M is the average cost of supplying the lost Btus, $/10^6$ Btu.

expanding

$$m_p = \frac{.524 \, k_1 k_2 \, \Delta T \, r_3 \, YM \times 10^{-6}}{k_2 r_3 \ln \frac{r_2}{r_1} + k_1 r_3 \ln \frac{r_3}{r_2} + k_1 k_2 R_s}$$

Now let $D = .524 \, k_1 k_2 \, \Delta TYM \times 10^{-6}$ \qquad \qquad E-4

so that
$$m_p = \frac{Dr_3}{k_2 r_3 \ln \frac{r_2}{r_1} + k_1 r_3 \ln \frac{r_3}{r_2} + k_1 k_s R_s} \qquad \text{E-5}$$

In a manner similar to that used in Equations A-14 through A-20, the following equation is obtained

$$\frac{(k_2 r_3 \ln \frac{r_2}{r_1} + k_1 r_3 \ln \frac{r_3}{r_2} + k_1 k_2 R_s)^2}{k_1 r_3 - k_1 k_2 R_s} =$$

$$\frac{D}{1.1 \, m_c B} = Z \qquad \text{E-6}$$

where D is the same as found in E-4
m_c is the installed incremental cost of insulation for the outer layer.

This is the equation which fairly defines the economic thickness of the outer layer given the already noted k value assumptions. To find r_3 using this exception the user is advised to first find Z from the steps outlined in Chapter 5, being careful to substitute $k_1 \times k_2$ for the k value in D_s or D_p.

Then choose a series of one-inch increments of r_3, starting from the value r_2 and calculating a series of values for the left hand side of E-6. Plot r_3 vs. Z_p.

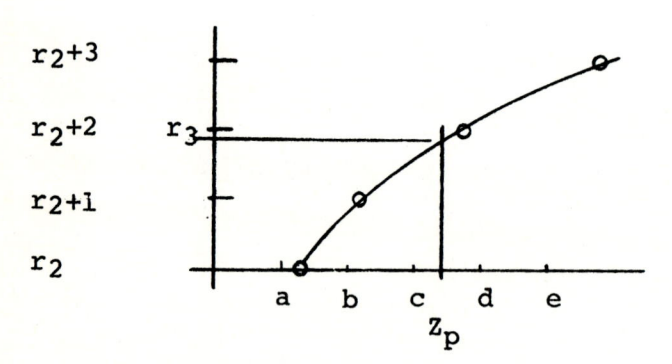

Figure E-2. <u>Means of finding R3</u>

Where the values of the left hand side of E-6 intersect the value of Z_p as determined from Fig. 5-7, the value of r_3 that would give the Z_p value has been determined. Then subtract r_2 from r_3 for the economic thickness of the second layer.

The use of E-6 is avoided in the manual by merely providing an average k value and solving as though the materials in the two layers were uniform. This leads to a higher degree of inaccuracy than E-6 provides, but by solving a number of sample problems it was found that an insulation thickness so chosen is usually within 1/2 inch of a more precise value.

Flat Surface

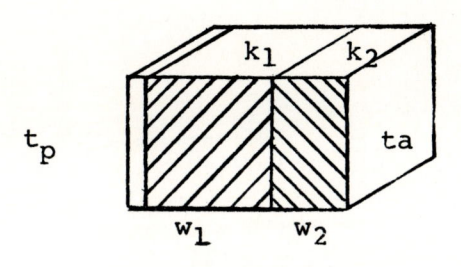

Figure E-3. Flat surface with two insulation layers

We have

$$\frac{q_s}{A} = U_s = \frac{\Delta T}{\frac{w_1}{k_1} + \frac{w_2}{k_2} + R_s}$$

as the heat flow equation with two thicknesses of dissimilar insulation.

Rearrange

$$U_s = \frac{k_1 k_2 \Delta T}{k_2 w_1 + k_1 w_2 + k_1 k_2 R_s}$$

multiplying in the cost of heat

$$m_s = U_s \, YM \times 10^{-6} \quad \frac{\$}{yr\text{-}ft^2}$$

expanding

$$m_s = \frac{k_1 k_2 \Delta TYM \times 10^{-6}}{k_2 w_1 + k_1 w_2 + k_1 k_2 R_s} \qquad \frac{\$}{yr\text{-}ft^2} \qquad \text{E-8}$$

The assumptions re k apply as with pipe.

Let $D_s = k_1 k_2 \Delta TYM \times 10^{-6}$ \hfill E-9

so that $m_s = \dfrac{D_s}{k_2 w_1 + k_1 w_2 + k_1 k_2 R_s}$

Then as before we end up with

$$(k_2 w_1 + k_1 w_2 + k_1 k_2 R_s)^2 = \frac{D}{1.1 m_c B} = Z_s \qquad \text{E-10}$$

Z_s can be found being careful to substitute $k_1 \times k_2$ for k in D_s and considering m_c only for the additional layer of insulation. And, as with pipe we can solve

$$(k_2 w_1 + k_1 w_2 + k_1 k_2 R_s)^2 \text{ for various values of}$$

w_2 until we find a w_2 which causes this value to equal Z_s.

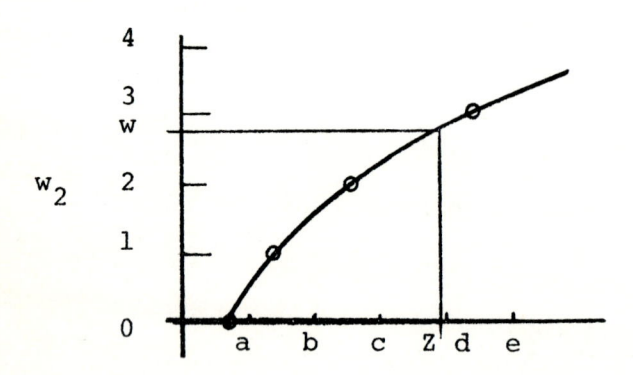

Figure E-4. Solution for least cost new insulation layer over existing layer

The above solution is not perfect, but is accurate enough for the economic thickness solution. As with pipe, an iterative solution for the proper k values may be made, but that solution is not required unless one is concerned about surface temperature, which is beyond the scope of this manual.

Using an average k value and solving the problem as in Chapter 7 is a further simplification, but it will also yield adequately accurate results.

APPENDIX F: PIPING COMPLEXITY FACTORS

Pipe fittings cost more to insulate than straight lengths of pipe. This higher cost is usually expressed in terms of "equivalent linear feet." For instance, if it costs $12 to insulate a pair of flanges connecting two lengths of pipe, and pipe costs $2/linear foot to insulate, the flange pair has an equivalent length of 6 feet.

Pipe fittings usually have a large surface area through which heat can be transferred as compared to a straight foot of pipe. Therefore, fittings have an "equivalent heat loss," also expressed in lengths of straight pipe. Usually the equivalent cost for insulating a fitting is greater than the equivalent heat loss area.

Since economic insulation thickness for pipe is a function of the insulation cost per unit of heat loss area, this relationship must also be true for a piping section, including fittings. Knowing this relationship, a factor can be derived for correcting unit insulation prices for pipe to include the effect of fittings. With this factor the economic insulation thickness for a piping section can be estimated based upon unit insulation prices for pipe.

For example: a line has 85 feet of straight pipe and a number of fittings, which cost the equivalent of 65 linear feet to insulate (total cost of 150 equivalent linear feet). The cost per straight linear foot for 1-inch of insulation is $4. The fittings have a heat loss area equivalent to 35 linear feet (total heat loss area of 120 equivalent linear feet). Therefore, the actual cost of insulation per equivalent foot of heat loss area is equal to 150/120, or 1.25 times the cost for a straight foot of pipe. The piping complexity factors listed in Table 4-1 are derived in the above manner and, therefore, enable the manual user to adjust the cost of installed insulation to account for fittings.

In evaluating piping complexity factors the number and types of fittings commonly found on piping must be known. Table F-1 lists the number and type of fittings typical of complex piping networks around process heat exchangers and equipment. For utilities and piping outside of process limits (typical of piping straight runs), approximately 1/3 the number of fittings shown in Table F-1 are likely to be found.*

---

*Data supplied by survey of Industrial Design Engineers.

Table F-1. Average Number of Fittings per 100 Linear Feet Pipe for Welded Piping

(Piping complexity is estimated as follows:

½ to 1½ inch, 1 fitting every 2½ feet; 2 to 6 inches, 1 fitting every 3 feet; 8 to 12 inches, 1 fitting every 4 feet; 14 inches and above, 1 fitting every 5 feet.)

Pipe Size (inches)	Flanges	Valves	Elbows	Tees	Reducers
½	11	5	7	2	1
3/4	12.5	6	18	2	2
1	18.5	9	17	4	6
1½	10	5	15	4	3
2	9	4	11	4	2
3	9	4	12	4	2
4	9	4	12	4	2
6	8.5	4	10	6	1
8	9	4	10	6	1
10	9	4	10	4	2
12	8.5	4	9	2	2
14	8.5	4	9	2	2
16-24	8	3	6	2	2

Note. Flanges are defined as line and valve flange mating pairs. The approximate costs for insulating fitting types, expressed as a multiple of pipe insulation cost, shown in Table F-2. These factors are based on data supplied by the National Insulation Contractors Association.

Table F-2. **Equivalent Length Cost Factor for Pipe Fittings**

Type	Pipe Size, inches			
	½"-6"	8"-12"	14"-18"	18"-above
Flange Pair	6	6	8	10
Valves				
Flanged	8	9	12	15
Screwed or Welded	2	3	4	5
Elbows				
Flanged	6	6	7	8
Screwed or Welded	2	3	4	5
Tees				
Flanged	8	9	10	12
Screwed or Welded	2	3	4	5
Reducers	1.5	1.5	2.0	2.0
Bent Pipe*	2	2	2	2
Hanger or Supports	2	2	2	2

*Denotes 2 times the developed length of pipe.

Note. Average heat loss areas for fittings expressed in equivalent feet of pipe are shown in Table F-3. These factors are based upon outside surface area of prefabricated 1½ inch thickness insulation covers applied to 150# schedule pipe and fittings. (The relationship of these fitting areas does not change appreciably with insulation thickness.)

Table F-3. Approximate Outside Surface Areas of Fitting Insulation (1½-inch thickness) Expressed in Equivalent Feet of Pipe Insulation

Pipe Size (inches)	Flange (Pair)	Valve	Tee	Elbow (90°)
½	1.33	2.86/1.14*	2.57/0.95	2.38/1.0
1	1.63	3.73/1.78	3.22/1.3	2.71/1.1
2	1.78	3.83/1.99	4.04/1.23	2.95/1.3
4	1.9	4.95/3.05	4.9 /1.5	3.95/1.6
6	1.83	5.08/3.25	5.31/1.75	4.52/1.74
8	1.75	5.45/3.7	5.22/2.0	3.76/1.94
12	1.84	5.94/4.1	6.01/2.74	4.8 /2.43
14	2.07	5.84/3.78	7/2.88	5.71/3.10
20	2.19	6.98/4.78	8/6.34	6.58/4.34

*Use top factor for flanged fitting and bottom factor for screwed or welded fittings.

Note. The data found in Table F-1 through F-3 were used to estimate piping complexity factors for both in-process and outside-process piping. These factors are shown in Table 4-1. In each case, the estimated total cost of insulating a 100-foot piping section (pipe and fitting insulation cost expressed in equivalent feet of pipe) was divided by the estimate total heat loss area (pipe and fitting areas expressed in equivalent feet of pipe) to obtain the complexity factor.